Palgrave Studies in Disability and International Development

We are pleased to announce the new book series, the Palgrave Studies in Disability and International Development. With this series, we open space for innovative research, debate and critical writings aimed at pushing forward the frontiers of discourse, theory and practice. We are seeking strong new monographs reporting on empirical work, edited books, as well as shorter theoretical writings, and are especially interested in interdisciplinary offerings. In the spirit of openness, we welcome unsolicited book proposals. We accept completed manuscripts, but would also be happy to hear on current research or on writing projects still in-process.

The series is intended to span a range of areas and we would welcome proposals on any topic related to international development and disability, including, though not exclusively:

- Inclusive education
- Employment and livelihoods
- Social protection Disability and poverty
- Human rights and disability rights
- Health and health care
- Discrimination and exclusion
- Religion and spirituality
- Disability definition and measurement—(Data and Disability)
- Rehabilitation and community based rehabilitation
- Enabling and disabling environments
- International development programs and their impacts on disabled people
- Disability cultures and identities
- Histories of disability
- Postcolonial issues
- Indigenous concerns
- Inclusive research and decolonizing approaches

More information about this series at
https://link.springer.com/bookseries/14633

Maya Kalyanpur

Development, Education and Learning Disability in India

palgrave
macmillan

Maya Kalyanpur
University of San Diego
San Diego, CA, USA

Palgrave Studies in Disability and International Development
ISBN 978-3-030-83988-8 ISBN 978-3-030-83989-5 (eBook)
https://doi.org/10.1007/978-3-030-83989-5

Cover image: Elena Ray/Alamy Stock Photo

This Palgrave Macmillan imprint is published by the registered company Springer Nature Switzerland AG
The registered company address is: Gewerbestrasse 11, 6330 Cham, Switzerland

I dedicate this book to Vijay Rao, who, in making growing old together fun, has made old age less inconvenient; to David Rao, who, in making his growing up fun for me, made unconditional love easier; to Sona Kalyanpur Rao, my north star; and to Laurel Carothers, the newest star in my firmament. In memory of Vasanti Savkur, whose companionship and conversations made two summer monsoons in Bombay easier to get through and brought back many memories of earlier childhood summers.

Foreword: Dislocating the Epistemic & Material Violence of Learning Disabilities in the global South

Maya Kalyanpur's *Development, education and learning disability in India* constitutes a significant intervention to dislocate the idea of learning disabilities (LD) in the global South.[1] This disability is a relatively new addition in the toolkit of classifications available in India to (purportedly) enhance educational opportunities for struggling learners, though Kalyanpur's research shows how the category contributes to ability difference stratification. Kalyanpur's study suggests that the adoption of LD in the Indian educational infrastructure must be examined through the historical prism of international development efforts and the global inclusive education agenda that has been exported from the US and Western Europe since the mid-1990s. The global emergence of inclusive education is characterized by at least four moves:

1. There has been an increasingly narrow focus on the education of learners with disabilities.
2. Although inclusive education policies have rapidly spread across geographical regions, the international development sector has slowly taken up inclusive education and disability, often in limited ways (Singal et al., 2019).
3. Conceptual ambiguities about inclusive education and implementation complications have emerged in nations of the global South (Grech, 2016; Werning et al., 2016).

4. The global spread of inclusive education and LDs has been fueled by a benign stance that strips these concepts of historical and cultural considerations (Artiles, 2019a).

Kalyanpur's research interrogates the convoluted global trajectories of inclusive education. She draws unsettling comparative parallels between the notion of LDs in the US and India. LD is the largest group of students in the special education population in the US and "are [considered] intrinsic to the individual, presumed to be due to central nervous system dysfunction, and may occur across the lifespan" (NJCLD, 2016). Moreover, the US definition deters intersectional analyses because environmental, cultural or economic factors are not considered causes of LDs (Tefera & Artiles, in press). Notwithstanding the largely technical approach and color-evasive stance used in the US, there is a tortuous record of entanglements with racial and linguistic differences in the history of LD (Artiles, 2019b, Donovan & Cross, 2002).

Despite the conceptual limits of the LD notion, Kalyanpur's study shows how India is swiftly moving to adopt the US model, starting with passing policies in 2009 that bestowed rights to this population and prescribed technical procedures. Many nations of the global South have followed this path not only for learners with LD, but in the broader arena of inclusive education (Artiles, 2019c). These moves constitute a form of *judicial deference*—i.e., the assumption that the mere presence of policies signifies that schools comply with the law and address the needs of the target student population (Edelman & Talesh, 2011). It is as if granting rights to the LD population will solve magically the multidimensional problem of educational opportunity and exclusion. Moreover, passing such policies is expected to signal advances in the development of the nation, in this case, in the education sector.

Kalyanpur uses an interdisciplinary framework to study LD in India that integrates postcolonial theory, Disability Studies and Disability Critical Race Theory. This is a welcome perspective urgently needed in the global South for it affords attention to historical, societal and cultural dimensions of disability. Her framework also accounts for the role of power in the construction and control of human categories such as disability, and it illuminates the analysis of LD in the context of colonial and imperialistic legacies in India. She asks, *"Why is the label itself based on the American definition? Why does the emergence of this category look similar to that in the US?"* Kalyanpur's study charts how the US framework and knowledge about LD is colonizing India's educational system. The Indian LD definition is a mirror reflection of the unstable US version. US assessment tools are prescribed to diagnose LD in India even though the tools lack appropriate cultural or linguistic norms. Even US terms like "high incidence disabilities" are being used in India despite the relatively small size of this population. The de-contextualized transfer of US LD knowledge to India constitutes a form of epistemic violence that ignores local contexts and silences local knowledges, particularly of the majority of the population that has been historically marginalized (Spivak, 1998).

Kalyanpur explains the uncritical adoption of the LD concept and technologies from the US constitutes a form of cultural hegemony that perpetuates the tyranny of the universal learner—i.e., the insistence on a monolithic human being in worlds of multiplying differences (Artiles, 2019c; Artiles & Caballeros, 2020). Moreover, Kalyanpur broadens the analytic scope by situating the introduction of LD knowledge from the global North in the contexts of globalization and neoliberal economic reforms in the global South. She reminds us neoliberal transformations have deepened and perpetuated social inequalities in India and other regions of the global South (Grech, 2015). A neoliberal ideology discourages social investments and privileges individual economic productivity. Considering that disability is one of the most powerful dimensions of exclusion (Stiker, 2009), the compounding of cultural hegemony in the context of globalization and neoliberal structures does not bode well for the potential of LD to advance an educational equity agenda in India.

In addition to the comparative layer of the study, a significant strength of Kalyanpur's research is the *situated* examination of LDs, particularly in the global South. A situated perspective requires researchers conduct their analyses with an eye on the global-local nexus (Goodley & Swartz, 2016) to illuminate "where and when disability appears and how it (re)emerges geopolitically, temporally and epistemologically" (Grech & Soldatic, 2016, p. 15). Kalyanpur juxtaposes the US-imported LD knowledge with the global South's multifaceted realities that inhabit India—a multilingual nation (22 official regional and 122 main languages) with two thirds of the population living in rural areas, and one-third oppressed by poverty. Of significance, a caste system is embroiled in socio-economic and linguistic structures, with *Dalits* enduring the economic and political brunt of the oppression. It is important to note, however, that Kalyanpur traced the historical origins of caste to racial differences—a fact that enhances the value of a comparative approach to study LD in the US and India. We also learn that the ripples of the caste system reach classrooms in the form of prejudice and other forms of violence.

In addition, a new form of language domination has emerged in India, in part due to the neoliberal impulse—indeed, English-speaking subjects have become a new caste due to the social mobility potential afforded by the language, though boundaries are increasingly re-drawn to preserve its elitist status. To illustrate, it is not enough to speak English, what *type* and *accent* of English one speaks matter. Kalyanpur contextualizes this evolving hierarchy-making process by noting that only 15 percent of the Indian population spoke English in 2019.

Given these circumstances, Kalyanpur sets to understand *who are the children with LD in India? What benefits do they accrue from this label? Are services differentiated by class/caste or language difference, as in the US*? In answering these questions, Kalyanpur documents how the epistemic violence of LD materializes to subvert LD's promise of educational opportunity in India. First, teaching and learning conditions are far from adequate in Indian schools due to large class sizes, underprepared teachers, under-funded schools and extraordinary demands on teachers to be responsive to constellations of socioeconomic, linguistic and caste diversity never imagined in the global North. Instruction in public schools is conducted in regional or national (Hindi, English) languages. It is not uncommon for children to speak local and regional languages, though in an example of neoliberal expansive reach, growing numbers of low-income students are enrolling in mix-quality private schools that pledge to teach English. This creates linguistic gaps to access the curriculum and contributes to depress academic achievement. Academic underperformance is extensive, but what proportion of school low achievement is shaped by LDs, lack of opportunity and/or structural conditions? The scarcity of qualified staff to make assessment and diagnostic decisions and the reliance on tests that are not normed in India complicate an LD explanation. Similar to the US, Kalyanpur documents a trend to diagnose students from the lower societal strata due to caste, language and socio economic differences. These cross-cultural resemblances deserve closer scrutiny to refine our understanding of the production of *DefectCraft*—i.e., the *othering* practices used in research, policy and practice to construct and perpetuate defect-laden identities primarily defined by individual traits, thus mystifying the cultural-historical mechanisms of power that mediate the stratification of ability and competence (Artiles, 2019b).

A relevant example is caste, linguistic or social class disparities in LD identification in India. *DefectCraft* assumes that disability diagnosis is the result of neutral practices and it is explained by psycho-biological deficits. *DefectCraft* dismisses intersectional analysis—e.g., caste-ability-social class-language intersections. Instead, *DefectCraft* focuses on a single identity vector (i.e., ability), thus making the role of caste or language *appear* or *disappear* at various key moments or situations in the disability identification process to legitimize the adscription of defects. Meanwhile, these practices obscure the role of historical barriers that reduce opportunities and produce linguistic and caste differences (Artiles, 2019b). A way to begin disrupting the work of *DefectCraft* is to examine not only the social *positions* of LD individuals (e.g., across intersections with language, social class, race), but also their social *positioning*—i.e., "how individuals come to be 'raced,' 'classed' and 'gendered' in relation to [*DefectCraft*] processes" (Morrison, 2012, p. 135, emphasis in original).

Of significance, the dense configuration of differences in the Indian educational context requires new forms of intersectional analysis. Intersectional analyses must transcend the traditional focus on two vectors of difference—e.g., LD + caste or LD + language. Instead, new framings should examine the dynamics of intersectional identities across types of settings (e.g., rural, urban; areas of conflict or disaster zones) (Artiles & Caballeros, 2020). For instance, how do the unique conditions of India's rural areas *open* and *close* opportunities for LD individuals? Beyond caste, social class, gender and language, what other identities and roles are available to LD individuals across settings? What are the consequences of these forms of representation for LD individuals? (Berghs & Kabbara, 2016) In other words, Kalyanpur's research challenges us to study identity intersections in the lived experiences of LD people (Morrison, 2012).

Ultimately, diagnosed students in India will find out that LD services are scarce in ordinary schools and special education schools are expensive. That is, learners may be assessed and labeled, but often do not get services due to high costs or lack of access to services. There is a perverse logic underlying this state of affairs: Ways to make up new kinds of children (i.e., LD) are imported from the global North without regard for local histories or epistemologies. Then, this form of epistemic violence materializes in the heterotopic spaces of Indian schools through naming new forms of learning needs (i.e., LD) followed by institutional inertia. In the end, Kalyanpur's study provides evidence that disability can be used as a justification to discriminate other marginalized groups (Baynton, 2001). Equally important, this work poses profound questions about "what counts as knowledge—how we know, what we know, why we know" (Behar, 2012, p. 208) and challenges us to forge new discourses and practices to engage with the state of affairs we call disability (Stiker, 2009).

Alfredo J. Artiles
Stanford University
Stanford, USA

Note

1. I assume the term global South signifies "power, resource, epistemological and other differentials, and though not unshiftable, and though not localized, embrace a substantial portion of the world living in a scenario of profound geopolitical asymmetries, poverty and isolation confronting deeply entrenched centers of concentrated wealth and power accumulated historically and perpetuated in times of coloniality. There is not one global South, but there are indeed many global 'Souths.' Furthermore, the global South is not only present in, but it also lives within the global north" (Grech, 2015, p. 15).

References

Artiles, A. J. (2019a). Inquietações sobre a educação inclusive no sul global: Um ponto de vista histórico-cultural. *Currículo sem Fronteiras, 19*(3), 1175–1187. [Published in Portuguese—*Probing inclusive education in the global South: A cultural-historical standpoint*] [ISSN 1645-1384 (online)]. http://www.curriculosemfronteiras.org/vol19iss3articles/artiles.html. http://dx.doi.org/10.35786/1645-1384.v19.n3.22.

Artiles, A. J. (2019b). Re-envisioning equity research: Disability identification disparities as a case in point. *Educational Researcher, 48*, 325–335.

Artiles, A. J. (2019c, July). *Disrupting inclusive education assumptions: Implications for the global South*. Paper presented at the biennial meeting of the International Association of Special Education, Sebastian Kolowa Memorial University (SEKOMU), Magamba, Tanzania.

Artiles, A. J., & Caballeros, M. Z. (2020, March). *Shades of rights and the dual nature of disability: Toward a cultural understanding of inclusion*. Highlighted session. Annual meeting of the Comparative & International Education Society. Miami, FL (Virtual session).

Baynton, D. (2001). Disability and the justification of inequality in American history. In P. K. Longmore & L. Umansky (Eds.), *The new disability history* (pp. 33–57). NYU Press.

Behar, R. (2012). What Renato Rosaldo gave us. *Aztlán, 37*(1), 205–211.

Berghs, M., & Kabbara, N. (2016). Disabled people in conflicts and wars. In S. Grech & K. Soldatic (Eds.), *Disability in the global South: The critical handbook*. Springer.

Donovan, S., & Cross, C. (Eds.). (2002). *Minority students in special and gifted education*. National Academy Press.

Edelman, L. B., & Talesh, S. A. (2011). To comply or not to comply—That isn't the question: How organizations construct the meaning of compliance. In C. Parker & V. L. Nielsen (Eds.), *Explaining compliance* (pp. 103–122). Edward Elgar.

Goodley, D., & Swartz, L. (2016). The place of disability. In S. Grech & K. Soldatic (Eds.), *Disability in the global South: The critical handbook*. Springer.

Grech, S. (2015). *Disability and poverty in the global South: Renegotiating development in Guatemala*. Palgrave Macmillan.

Grech, S. (2016). Disability and development: Critical connections, gaps and contradictions. In S. Grech & K. Soldatic (Eds.), *Disability in the global South: The critical handbook*. Springer.

Grech, S., & Soldatic, K. (2016). Disability in the global South. In S. Grech & K. Soldatic (Eds.), *Disability in the global South: The critical handbook*. Springer.

Morrison, A. (2012). If you are holding a degree in this country no-one's gonna ask no questions. In K. Bhopal & J. Preston (Eds.), *Intersectionality and "race' in education*. Routledge.

National Joint Committee on Learning Disabilities (NJCLD). (2016). *Definition of learning disabilities*. http://www.ldonline.org/pdfs/njcld/NJCLDDefinitionofLD_2016.pdf.

Singal, N., Lynch, P., & Johansson, S. T. (Eds.). (2019). *Education and disability in the Global South: New perspectives from Asia and Africa*. Bloomsbury.

Spivak, G. (1998). Can the subaltern speak? In C. Nelson & L. Grossberg (Eds.), *Marxism and the interpretation of culture*. University of Illinois Press.

Stiker, H. J. (2009). *A history of disability*. University of Michigan Press.

Tefera, A., & Artiles, A. J. (In press). Learning disabilities' unsettling intersections: Betwixt learning, cultural, socioeconomic, and environmental deficiencies. In R. Tierney, F. Rizvi, K. Ercikan, & G. Smith (Eds.), *International Encyclopedia of Education* (4th ed.). Elsevier.

Werning, R., Artiles, A., Engelbrecht, P., Caballeros, M. Z., Hummel, M., & Rothe, A. (Eds.). (2016). *Keeping the promise? Contextualizing inclusive education in developing countries*. Klinkhardt. http://www.pedocs.de/frontdoor.php?source_opus=12353&la=de

World Health Organization & World Bank. (2011). *World report on disability*. Switzerland.

Acknowledgments

I wish to thank Sandip and Pratyush Hire, for their generosity and willingness to share their time, their experiences and their home with me; Dr. Sandhya Limaye, Professor, Centre for Disability Studies and Action, Tata Institute of Social Sciences, who made the research possible with her immense network of connections; Radhika Misquitta, Co-Principal at The Gateway School of Mumbai, who taught me so much about the field of learning disabilities in India; Kate Currawala, Co-Founder, and Masarrat Khan, Maharashtra Dyslexia Association, Dr. Nandini Singh, then at the National Brain Research Center, and Dilip Rangnekar, CEO, Azim Premji Foundation, who shared their experiences and expertise on the intricacies of assessment and instruction, language and learning and the broader educational system in India; and Ankit Shah, doctoral student and research assistant, who provided invaluable assistance with the manuscript, with his firsthand knowledge of the Indian educational system and his facility with Hindi for the translation of documents. Heartfelt gratitude to the schools in this study and, particularly, the teachers, the students and their parents. I also wish to acknowledge the support of the Fulbright Foundation through a senior research fellowship that made this study possible.

Praise for *Development, Education and Learning Disability in India*

"This insightful book offers a critical discussion of disability and education in India that highlights the historical, political, sociocultural, economic, ideological and other local and global factors that shape inclusion and exclusion in this vast multilingual and multicultural global South country. Underpinned by a colonial/post-colonial and disability studies theoretical framework and based on a study conducted in 6 Mumbai schools, the book illustrates the complex intersections that create and perpetuate disadvantage, marginalisation and exclusion in the Indian context, as well as the role of discourses and practices uncritically imported from the North. The author's particular position as an insider/outsider in relation to the Indian context gives the book a unique lens, which sheds light on the tensions of translating global education agendas into local contexts. Written in an engaging and accessible style, this book is a must-read for anyone interested in inclusive education development, the intersections between disability and other forms of disadvantage, decolonising methodologies and social justice."

—Dr. Leda Kamenopoulou, *Senior Lecturer in Special and Inclusive Education, School of Education, University of Roehampton, UK*

Contents

About the Author

Maya Kalyanpur, Ph.D., is Professor in the Department of Teaching and Learning at the University of San Diego having received her Ph.D. in Special Education from Syracuse University. She started her career as a teacher of children with intellectual disabilities in India. She was an international advisor in inclusive education to the Cambodian Ministry of Education under the World Bank-supported Global Partnership for Education program and received a Fulbright senior research fellowship to study services for students with learning disabilities in India. In 2020, she was awarded the American Educational Research Association (AERA) Multicultural/Multiethnic Education Carlos J. Vallejo Memorial Award for Lifetime Scholarship, and was Cognizant Visiting Professor and Chair at the Center of Excellence in Teacher Education, Tata Institute of Social Sciences, Mumbai, India. She has conducted research on the intersections of culture and special education and international development and disability studies in India, the US and Cambodia, with a specific focus on international inclusive education policy and practice, and families of children with disabilities from culturally and linguistically diverse backgrounds. She co-edited the book, *Disability studies in South Asia: Redefining boundaries and extending horizons*, with Dr. Shridevi Rao, which presents an alternative analysis to disability appropriate to South Asia, and is currently co-authoring a book on English language education and social inequality. She co-founded the Proteep project, which supports girls to complete high school in Cambodia.

Abbreviations

ASER	Annual State of Education Report
BCE	Before the Common Era
BMC	Bombay (Mumbai) Municipal Corporation schools
CARS	Childhood Autism Rating Scale
CBCS	Center of Behavioral and Cognitive Sciences
CBSE	Central Board of Secondary Education
CCE	Continuous & Comprehensive Evaluation policy
CES	Center for Equity Studies
COVID-19	Coronavirus Disease 2019
CRPD	(UN) Convention on the Rights for People with Disabilities
CSR	Corporate Social Responsibility Act
DALI	Dyslexia Assessment for Languages of India
DIBELS/ORF	Dynamic Indicators of Basic Early Literacy Skills/Oral Reading Fluency
DIKSHA	Digital Infrastructure for Knowledge Sharing
EFA	Education For All
GDP	Gross Domestic Product
GRR	Gradual Release of Responsibility
IB	International Baccalaureate program
ICDS	Integrated Child Development Services
ICSE	Indian Council of Secondary Education
IDEA	Individuals with Disabilities Education Act (US)
IEDC	Integrated Education for Disabled Children
IELTS	International English Language Testing System
IGCSE	International General Certificate of Secondary Education exam
IMF	International Monetary Fund

LFP	low-fee-paying private schools
MDA	Maharashtra Dyslexia Association
MHRD	Ministry of Human Resource Development, now the Ministry of Education
MOI	Medium of Instruction
NASEOH	National Society for Equal Opportunities for the Handicapped
NCERT	National Council of Educational Research and Training
NCTE	National Council for Teacher Education
NDP	No Detention Policy, also known sometimes as the No Retention Policy
NEP	National Education Policy
NGO	Non-Governmental Organization
NIOS	National Institute for Open School
NJCLD	National Joint Committee on Learning Disabilities (US)
OBE	Open School Board
OT/PT	Occupational and Physical Therapy Services
PUC	Pre-University College
RCI	Rehabilitation Council of India
RPD	Rights of Persons with Disabilities Act, formerly known as the Persons with Disabilities (PWD) Act
RTE	Right to Education Act
RTI	Response to Intervention
SC/OBC	Scheduled Castes or Other Backward Classes
SLD	Specific Learning Disabilities
SSA	Sarva Shiksha Abhiyan, also Samagra Shiksha Abhiyan
SSC	Secondary School Certificate
TLF	Three Language Formula
UDL	Universal Design for Learning
UN	United Nations
UNESCO	United Nations Educational, Scientific and Cultural Organization
UNICEF	United Nations Children's Fund
WHO	World Health Organization
WISC	Wechsler Intelligence Scale for Children

CHAPTER 1

The Emergence of the Category of Learning Disability in the US and India

At first glance, the US and India would appear to have little in common, as I believed when I first came to the US from India as a graduate student. Yet, over the years, I have come to see similarities. Both were British colonies, America's revolution drawing Britain's attention to India, and are currently democracies. India's caste system endures through the ages, while America's legacy of slavery has created a similar caste system of social and economic inequality (Wilkerson, 2020). But many similarities are not coincidental. In our current postcolonial era of globalization and international development, there is a deliberate stance that has India looking toward the US for direction and following a beat behind in ideology, policy and practice. The neoliberal market fundamentalism that has created enormous wealth inequalities in the US is reflected in the post-1990s Indian economic-political strategy. The upper and middle class in India is now just as familiar with American television celebrities and shows like Sally Jesse Raphael, Oprah, *Baywatch* and *Friends* and celebrates Valentine's Day, Mother's Day and Father's Day with equal enthusiasm. Elite schools prepare students with skills such as critical thinking, social emotional learning and co-curricular social activism which echo pedagogic principles in the US, often with the explicit purpose of helping them get admission into American colleges and universities. And, just like in the US, despite the intention of making schooling universally accessible and

M. Kalyanpur, *Development, Education and Learning Disability in India*, Palgrave Studies in Disability and International Development,
https://doi.org/10.1007/978-3-030-83989-5_1

including traditionally excluded children, students are being marginalized. In particular, following the US trajectory, a new group of students is being labeled learning disabled (LD) and inequitably experiencing the benefits or disadvantages of this label based on their socio-economic status. This is the focus of this book.

This book foregrounds the theoretical frameworks of postcolonialism and disability studies, which share a pivotal imbrication of the process of "othering," or the creation of a perception of difference that is mostly deficit-based (Kliewer & Fitzgerald, 2001; Simpson, 2007). Applying these lenses, it explores the implications of this west-looking stance resulting from neoliberal globalization with regard to the emergence of a new category of disability in India. It is counterintuitive that the global development agenda of Education For All (EFA) with its inclusionary aims of providing access to quality education to historically marginalized groups should have created a new perception of difference and marginalization. Further, the over-representation of linguistically diverse and socio-economically disadvantaged children in the LD category in the US (see, e.g., Blanchett et al., 2009; Duhaney, 2014) may also have transferred to India, given its own multilingual and resource-limited context.

This chapter situates this analysis within the overarching theoretical frameworks of postcolonialism and disa bility studies, and within the nested theories of (a) disability and critical race theory (or Dis/Crit) which examines the intersectionality of disability with other disadvantaging factors, such as language and socio-economic status, and (b) disability studies in education, which explores these ideas within the specific context of educational systems. It compares the emergence of the LD category in the US with that in India as a springboard to the analysis presented in subsequent chapters, concluding with a description of the methodology of the study. But first, a note about my positionality.

My Positionality

The origins for the study probably began in the early 1980s when I was a teacher at a special education center in New Delhi, India, working with students with mild and moderate intellectual disabilities, whom my colleagues and I were trying to include in general education classes. Looking back, I realize some might have been re-labeled according to today's sensibilities, but the field was in its infancy in India at that time and many of us were looking to the US and UK for answers. Occasionally, my colleagues and I wondered if these young boys and girls would qualify as having LD, a term we were just hearing but knew little about. Hoping that being new and therefore more value-neutral, it might be less likely to provoke the outright rejection that intellectual disabilities did from general education teachers, we began to use the term to describe many of our students, particularly those that we saw as being "eligible" for inclusion. All my students were Hindi-speakers, and most spoke another regional language, like Punjabi, Urdu, Rajasthani, or Tamil at home. For many, English was a new language they were trying to acquire in school, with some difficulty. As a teacher, however, I rarely recognized their assets as multilingual speakers as such, because of how complicated instruction became in such a situation. I remember vividly trying to teach a lesson on prepositions, where I had them climbing on, in, under and over their desks and other spaces in the classrooms to convey the meaning of the terms. To facilitate their comprehension, I spoke in Hindi and provided the academic terms in English, only to realize that quite a few of my students were confusing the English word "under" or below with the Hindi word "*andar*" or in. At this point, I switched to the word "*bheetar*" in Hindi, a synonym for *in* that is more formal and therefore rarely used. I'm not even sure if the students knew the word. What a disastrous lesson for all!

Yet, for all their struggles in school, these students were smart as whips. On a field trip, one 12-year-old student, despite his thirst, refused to buy a cold drink from a street vendor because, he pointed out, the man was charging him three times the regular price "*aur wo bhi garam*!" ("and that too unchilled!"). He knew right away when he was being scammed. Although our attempts at inclusion were only partially successful, primarily because of the academic pressures of the English-medium curriculum, they would go on to make a success of their lives. One young man became a gentleman farmer, learning to oversee his family farm with the help of a bailiff who managed the financial aspects. Another became a cabinet-maker with his father. One young lady, with the help of her parents, set up and ran a beauty salon from their home.

But these successes came later. In the meantime, convinced that I needed "more strategies" as a teacher, I joined an American university naïvely hoping to find the right solutions to meet my students' needs in India. My first field placement was in a group home, where six adults with developmental disabilities lived together under constant supervision. My instructor told me that group homes gave the residents a sense of their independence as adults, but I wondered if group homes made sense in Indian society where even typical adult children did not leave home, and the milestone for independence and adulthood is typically marriage.

For a course assignment, I observed a student being tested on a standardized reading assessment. The 10-year-old boy was asked to read a list of words assumed to be part of his everyday vocabulary. He successfully read "spout" and "catch" but had trouble with "floor" and "rhyme." I was sure he would not be able to read the next word, but to my astonishment, he read it quite correctly and with triumph! The word was *jeopardy*. The popular American quiz show "Jeopardy!" had made it, in effect, an everyday word for him. I began to realize that the field of special education, including the assessment of "intelligence" and competence, is culturally and contextually defined.

Over the decades, the field of inclusive education has grown in India, as well as the scholarly work that acknowledges the need to seek indigenous solutions and/or adapt to the Indian context what the US has to offer. It is a delicate balance trying to determine what is appropriate to bring in and what needs to be adapted. The rising number of children who are being stigmatized by the LD label in India is a concern. In some cases, since the label provides access to services, these benefits are perceived to outweigh the negative consequences of the stigma. In many cases, the perception of disability is complicated by language difference, as happened with my students, or by class/caste disparities.

My insider/outsider status as an Indian living in the US gives me both disadvantages and advantages. As an outsider, I can see the bigger picture of the inexorable move toward development but I am also away from the day-to-day pressures of working as a teacher or inclusive education professional within India. As an insider who is multilingual like most Indians, I can understand the nuanced complications of language and medium of instruction in classrooms but I am also removed from the justification for pursuing these global aspirations. Most significantly, I am aware that I could be perceived as an outsider pointing fingers, which is certainly not my intention. From those early days when we were just becoming familiar with the term LD, the field has come a long way in India. A strong corpus of disability studies scholars and scholars in the LD field are acutely responsive to seeking context-specific solutions and foregrounding indigenous experiences and knowledge, and this book acknowledges this deep, emerging scholarship. Despite these efforts, however, the parallels with LD in the US are uncanny and we need to examine them more closely. Whom are we serving if, in trying to provide an education for all, we are actually creating a new group of marginalized students? Does access to and acquisition of English truly facilitate the social aspirations of low-income, linguistically diverse children and families or does it only further privilege the already advantaged? Have we set up educational structures that are contextualized for India and serve the needs of each child, or have we constructed systems that serve the larger global context of neoliberal economics in the name of development? Based on a study conducted in India, this book explores the unsettling resemblances in the development of LD in the US and India, the larger, global variables that have contributed to this development, and the troubling outcomes for children who are being labeled and stigmatized because they may not meet these arbitrary standards of competence or intelligence.

The next section describes the overarching theoretical frameworks of postcolonialism and disability studies, and the nested theories of disability and critical race theory (or Dis/Crit), which examines the intersectionality of disability with other disadvantaging factors, and disability studies in education, which explores these ideas within the specific context of educational systems.

Postcolonial and Disability Studies Lenses

This book is situated at the intersection of colonial/postcolonial studies and disability studies or "the symbiotic subjugation of people with disabilities and others who strayed in social appearance from the Western ideal" (Kliewer & Fitzgerald, 2001, p. 467). Scholars building on Foucaultian theory have argued that the perception of the "Other" created during colonialism to justify slavery and imperialism informed the perception of deviance and provided the rationale for the concept of the norm (Kliewer & Fitzgerald, 2001; Said, 1979; Simpson, 2007). The "gaze," or the act of seeing that defines a specific, usually negative, perception for the onlooker, for both imperial subjects of so-called other races and people with disabilities is similar in that it dehumanizes the person or persons being gazed at while allowing the gazer to engage in a sense of their own perceived superiority. As Gramsci notes, this perception of superiority is then translated into a cultural hegemony that assumes these perceptions or beliefs to be universal, and imposes them in ways that renders the oppressed unaware of being oppressed. The sense of "otherness" is situated in both perception and place (Ferri & Connor, 2006; Naraian, 2016; Ware, 2010). In our context, the othering manifests between what used to be the colonizer, now the developed world, or the global North and the colonized, now the developing world, or the global South. It also extends beyond the chronology of colonialism into a contemporary globalization as Empire (Hardt & Negri, 2000), where a country such as the US exerts its epistemological hegemony without actual physical annexation.

Based on the Cartesian notion ("I think, therefore I am") that knowledge is created by the perceiver, the hegemony of Empire assumes that "knowledge produced from the social/historical experiences and world views of the Global South, also known as 'non-Western', are considered inferior and not part of the canon of thought" (Grosfoguel, 2013, p. 75). Extending Said's (1979) colonial/postcolonial argument that the colonizer determines how the Orient is perceived, the othering in disability studies imposes a tyranny of the norm. Racism based primarily on skin color difference emerging as a direct outcome of colonialism, slavery and the Empire is mirrored in ableism or the discrimination based on differences in physical appearance and mental acuity (Kliewer & Fitzgerald, 2001; Simpson, 2007). In both cases, specific populations are categorized, deemed inferior and deviant, toward justifying their "confinement" (Kliewer & Fitzgerald, 2001) or segregation, for instance, through laws against racial miscegenation or institutionalizing special schools. These places of containment were called "colonies," perhaps not coincidentally,

> admit(ting) anyone, in short, whose behavior, desires, choices, or appearances fell outside the accepted norm.
>
> Feeble-minded women were sent to the Virginia State Colony for confinement to ensure that they would not continue breeding… The word *colony* gave its purpose away: the place was never to be a hospital or asylum. Rather, from its inception, it was designed to be a containment zone. (Mukherjee, 2016, p. 79, emphasis in original)

As the concepts of normalcy and, conversely, deviance took hold, processes for identifying difference became paramount, resulting in the development of the field of psychology and psychometric assessments such as IQ tests and cranial classification of the races (Gould, 1996; Parekh, 2017; Simpson, 2007). Even within our current, postcolonial world, many of these measures continue to be used universally to separate and to justify the segregation of populations deemed as the "other." Writing about the US, Kozleski (2020) asserts, "dis/ability has been conceptualized and understood as an individual deficit that precludes participation within the larger community. Additionally, it reminds us that dis/ability labels have been used to separate and confine individuals who do not conform to local social norms" (p. 343). Similarly, Maudslay (2014) notes, "In many ways Western professionals' understanding of Asian concepts of disability still appear trapped within an Orientalist framework (Said, 1979) in which Western perceptions imposed a set of generalised characteristics onto non-Western people" (p. 419).

Intersectionality and Disability Critical Race Theory

Scholars of disability critical race (Dis/Crit) theory have argued the need for more critical understandings of the intersection between disability, class, race, gender and sexuality (Annamma et al., 2018; Erevelles, 2011). This stance recognizes that the intersectionality of micro-cultural components of identity already perceived as disadvantaged can exacerbate the effects of discrimination. For instance, in a patriarchal and ableist world, women and girls with disabilities are doubly disadvantaged by being both female and disabled (UNICEF, 2013). The duplicative disadvantaging effect of poverty on disability, in that poor people are more likely to become disabled because of limited or lack of access to proper health care and disabled people are more likely to become poor because of their disability, is well-documented internationally (Arne & Ingstad, 2011; Grech, 2009). Similarly, transgender studies and gerontology intersect with disability studies in defying the notion of a normative corporeal body (Chennat, 2019a).

Much of the research in the recent but robust field of disability studies in India considers the issue of intersectionality (see Addlakha, 2013; Mehrotra, 2020a) within its broad multidisciplinary reach (Karna, 2001). For instance, scholars have provided nuanced analyses of femininity and masculinity within a patriarchal castcist and ableist society from Indian literature (Anand, 2015), personal experiences as a disabled woman (Ghai, 2020) and interviews with disabled men (Mehrotra & Vaidya, 2020). More significantly, this scholarship has made efforts to diverge from Northern constructions of disability and societal responses to disability as being embedded in colonialist discourse, arguing that, unlike in the west, charity and rehabilitation models have as much of a role to play as rights-based platforms (Mehrotra, 2020b) and that disability studies in South Asia must consider "experiences related to colonialism, caste and gender, as well as those shaped by the neoliberal rationalities that have become prominent since the early 1990s" (Staples, 2020, p. 26). Some have identified the intersections between Amartya Sen's capabilities approach (CA) as a humanistic model of economic theory and disability studies (Chennat, 2019b; Hammad & Singal, 2015; Mitra, 2006; Terzi, 2010). CA seeks to explain economic constructs such as poverty and quality of life through the nuances of lived experiences in the South, whereby individual differences in practical opportunity or access and freedom of choice define the level of a person's capability and functioning (Mitra, 2006). For instance, although the government may provide wheelchairs free of cost, a person's ability and freedom to participate in society may still be restricted by structural barriers such as limited architectural accessibility, affecting their quality of life in terms of their health and their self-respect (Chennat, 2019b).

Disability Studies in Education

Disability studies in education has sought to deconstruct the structural assumptions of educational systems that have contributed to the identification and marginalization of specific groups of children (Dudley-Marling, 2004). Several scholars have posited that educational systems reflect the social hierarchies of their societies (Bourdieu & Passeron, 1990; Bowles & Gintis, 2011; Foucault, 1972; Illich, 1971; Giroux, 1983). While, on the one hand, schooling is perceived to be the avenue for social mobility in providing access to knowledge, the epistemology favored in schools is often already accessible to upper classes, disadvantaging those who do not belong to the elite. As a result, schooling becomes a means of social reproduction (Bowles & Gintis, 2011; Collin & Apple, 2010). The "tyranny of ability" that becomes a stratifying device in classrooms lends itself within society to the myth of meritocracy and the belief that people have earned their privilege of status instead of acknowledging the privilege of what is essentially an accident of birth (Ladwig & McPherson, 2017; Parekh, 2017). Analyzing the French educational system, Bourdieu noted that the classical curriculum and the academic, standard French language taught in schools disadvantaged children from working-class backgrounds. As Foucault (1972) asserts:

> Education may well be, as of right, the instrument whereby every individual, in a society like our own, can gain access to any kind of discourse. But we well know that in its distribution, in what it permits and in what it prevents, it follows the well-trodden battle-lines of social conflict. Every educational system is a political means of maintaining or of modifying the appropriation of discourse, with the knowledge and the powers it carries with it. (p. 227)

As Naraian (2019) notes, scholarly critiques of the larger general educational system in India do not always consider students with disabilities as a constituency. The separation of the education and disability spaces, suggests Chennat (2019c), is an outcome of the perception that "inclusion in education is a separate phenomenon not in zinc (sic) with inclusion in society" (p. 54). This may simply be because, till recently, there were few children with disabilities in school (UNESCO, 2019; Vaidya, 2020), and attention was focused on the many other under-represented minority groups, such as the poor, girls and *Dalits* or lower castes (Kalyanpur, 2008a). Pioneering scholar Mithu Alur (2002) first raised the issue of the invisibility of children with disabilities in government programs, specifically the early intervention-focused Integrated Child Development Scheme (ICDS) targeting poor children. As government efforts to include children with disabilities have increased since the new millennium in response to international policies on inclusive education, research on the intersectionality of school and disability has been extensive and analytically critical. With children with disabilities being the last group to be included in education (UNESCO, 2019; World Health Organization/World Bank, 2011), scholars have noted the considerable progress made in legislation, policies and programs while also pointing to the many challenges and gaps that still persist (Jangira, 2002; Jha, 2004). These include attitudes toward disabilities, both societal and those of general education teachers reluctant to accommodate children with disabilities in already oversize classes (Ahsan et al., 2013; Limaye, 2019; Timmons & Alur, 2004), ambiguities in perceptions of the meaning of inclusion (Kalyanpur, 1996, 2008b; Naraian, 2016; Singal, 2005, 2006) and limited human and material resources.

The Emergence of Learning Disabilities

Scholars have argued that perceptions of deficit based on presuppositions of what is “normal,” with particular focus on classroom behavior and academic expectations, are often idiosyncratic, lacking a clearly defined etiology (e.g., Baglieri & Shapiro, 2017; Connor et al., 2015). The LD category is particularly troublesome in this regard, as this section shows through an analysis of its emergence in the US and India.

Learning Disabilities in the US

The definition of learning disabilities states that “these disorders are intrinsic to the individual, presumed to be due to central nervous system dysfunction, and may occur across the lifespan” (US National Joint Committee on Learning Disabilities [NJCLD], 2016). It adds that although extrinsic influences such as cultural differences, insufficient or inappropriate instruction may occur concomitantly, they are not the result of these influences. Scholarly literature situated ideologically within the medical model of intrinsic deficit has sought to understand why some students manifest difficulties in the acquisition and use of listening, speaking, reading, writing, reasoning or mathematical abilities, and to identify instructional and medical interventions that could mitigate, remediate or circumvent these difficulties in learning (see Thapa et al., 2008, for a review).

However, several US scholars have asserted that this category is a social construction (see Duhaney, 2014, for a review), a “soft disability,” because of the lack of definitive etiology, its definition focusing more on an elimination of factors not related to vision, hearing or motor impairments. In 1973, Jane Mercer coined the terms “the six-hour retarded child” or “the situationally retarded” for children placed in special education settings for academic performance or behavior deemed non-normative, even though within their home communities, which were largely Hispanic-American and African-American, these children were perceived as typical, meeting all the general expectations of children of their age in their society. Their families were unaware their children were being segregated in school. Mercer concluded that differing expectations of competence between home and school were resulting in students being perceived as deviant and therefore marginalized.

In 1986, Christine Sleeter, noting a meteoric increase in the numbers of students being labeled LD since the 1950s, argued that the space race between the US and the USSR in the 1950s contributed to an increased competitive academic environment within schools. This led to students who were not succeeding academically to be moved out of regular education and categorized LD. As Sleeter argued, the students did not necessarily have a specific innate medically evidenced cognitive disability, they were just unfortunate to be going to school at the time when more was expected of students, leading to the social construction of LD as a way of labeling students who were not making the grade.

In their personal accounts, Jonathan Mooney (2007) and Christopher Lee (Jackson & Lee, 1992) attribute their initial failure in school to the shortcomings of an educational system unable to offer flexible approaches to students and the embedded assumption of deficit related to experiencing difficulty in reading. As the world has moved increasingly toward the expectation that all children, all people, should be able to read, children who do not acquire this skill are stigmatized and marginalized and seen as somehow less competent. The fact is children are not born readers and need to be taught to read, and some children struggle in this process. Herein lies "the dilemma of difference" (Norwich, 1996, cited in Terzi, 2010, p. 36): To assume incompetence because some children may not have had exposure to or been taught these skills is as unfair as the assumption that the provision of remedial interventions outweighs the stigma of this perceived incompetence. Placing the problem within the child as an intrinsic deficit overlooks the structural inequities that contribute to the problem occurring in the first place and only results in blaming the victim. As McDermott et al. (2011) eloquently assert:

> Every disability is best understood as a situation rather than a personal trait. One cannot be disabled alone; it takes a village, it takes a state, it takes a capitalist democracy in which every person being less means that someone else is being more. And so the labeled are treated.

More disturbingly, there are class and race/ethnic differences in who gets labeled, resulting in a disproportionate representation of poor students, students of color and students for whom English is not their first language (Annamma et al., 2018; Blanchett et al., 2009; Erevelles, 2011; Gabel, 2004; Harry, 2014). Recent scholarship asserts that children who come from English-speaking, middle-class backgrounds, who are exposed to pre-literacy skills in preschool or at home, because parents have the resources to buy books, and the cultural expectation and requisite level of education to read to their children, are privileged when they enter school, since they already have many of the skills and competencies to become successful students who can learn to speak, read and write fluently in academic, standard English. Children who have not had these prior experiences start off disadvantaged, and many remain so. Their inability to become successful students has less to do with any innate deficits, but with social inequities of access and opportunity, and structural inequities in educational systems that privilege certain languages and knowledge. Although students with LD currently constitute over 50% or more than half the number of children with disabilities receiving special education services of all children with disabilities in the US, the controversy about this label persists: Is LD a condition intrinsic to the child or is it a constructed category based in societal perceptions and arbitrarily established expectations of student competencies?

Learning Disability in India

This question is equally relevant to the Indian context. For one, the label of LD is itself new to India, indicated in part by the fact that the local term for it is derived from formal Hindi (अध्ययन अक्षमता *addhyayan akshamataa*).[1] Secondly, there is evidence that the lens by which these children are being identified is imported from the US. For instance, the definition for the term in the Rights of Persons with Disabilities (RPD) Act bears a strong resemblance to that in the US Individuals with Disabilities Education Act (IDEA) (Unni, 2012). However, while drawing from research in the global North to explain the problems of children who struggle to learn to read, scholars also emphasize the importance of recognizing context-specific factors, such as bi-/multilingualism, far from ideal classroom conditions and socio-economic disadvantage (Das, 2008; Karanth, 2008; Pandey & Pant, 2008; Ramaa, 2000; Thapa, 2008). In a review of research on LD in the last two decades of the twentieth century, Ramaa (2000) notes that “conducting studies of learning disabilities in India is not an easy task because of its multilingual and multicultural background” (p. 268) and concludes that:

> In India identification and diagnosis of learning disabilities is very difficult because ready-made standardized tools are not available in all of India’s

> different languages. There is an urgent need to develop standardized, teacher-made assessment instruments in the different languages and to establish norms. (p. 271)

In her edited book, Komilla Thapa (2008) notes that, as students with reading difficulties began to be noticed from the beginning of the new millennium, there has been a steady increase in the "identification of children with LDs and a consequent demand for services" (p. 26). She describes the beginning of scholarly focus on the field of learning disabilities at the International Conference on Learning Disabilities and Young-At-Risk organized by the Center of Behavioral and Cognitive Sciences (CBCS) held in February 2003 in Allahabad. "Since research and practice in the area of learning disabilities (were) still at a nascent stage..., the papers focused on issues related to the conceptualization and definition, assessment and interventions of learning disabilities, the role of the teacher in identification, the salience of contextual factors and the language learning-learning disability connection in Indian languages" (Thapa, 2008, pp. 18–19). At this event, and in the book that emerged from it, scholars specifically cautioned against the use of the IQ-achievement discrepancy as in the US, asserting that LD be defined instead by deficit in cognitive processing (Das, 2008; Thapa, 2008). As Thapa (2008) stated:

> In India, we should not give the classical IQ testing based on WISC and Binet (translated or in original) any importance in determining learning or reading disability; thus, we will avoid the mistakes made in the past. An alternative to IQ testing is available in the form of Cognitive Assessment System especially as it works well with Indian children in English-medium schools. Moreover, the program is widely popular even using American norms. (p. 15)

In relation to the context-specific issue of bi/multilingualism, scholars also suggested that providing intensive phonics training and the Indian methods of writing to students who are already familiar with Indian languages might help to reduce the incidence of dyslexia in India (Balasubrahmanyam, 2001, cited in Thapa, 2008; Karanth, 2008).

Learning disabilities was officially recognized only as recently as 2009 when the Persons with Disabilities Act of 1995 was amended to include the category of Specific Learning Disabilities (Unni, 2012). Yet, since the new millennium, the number of children labeled as LD or dyslexic in India has increased exponentially (Karande et al., 2011) and in 2011, students with LD constituted 7% of all children with disabilities (Government of India Census, 2011, cited in UNESCO, 2019).[2,3] In 2012, about 10% or 30 million children were estimated to have LD ("10% of kids," 2012). One reason may be that, in postcolonial India, more children than ever before are accessing an education in the first place, facilitated by the Indian government's Samagra Shiksha Abhiyan[4] or EFA program launched in 2001, and the 2009 Right to Education (RTE) Act. Further, private school enrollment in most states is increasing (Annual Status of Education Report [ASER], 2019), particularly as parents, recognizing English as the language of opportunity and social mobility ("Goddess of English," 2011; Motha, 2014; Varma, 2007) and regardless of whether English is spoken at home (Kalia & Reese, 2009), are increasingly choosing to send their children to English-medium private schools. The medium of instruction in government schools is in the regional or national language, although this is changing. With the commensurate increase in competitiveness to get into and do well in school, students who experience academic difficulties are often perceived as "hopeless or badly behaved" and labeled LD ("Pain of Dyslexia," 2008). While it is presumed that the main benefit of labeling is that students will get remedial help, it can also be problematic (Terzi, 2010).

One, there is a stigma associated with disability (Center for Equity Studies [CES], 2014). In 2007, the blockbuster Bollywood movie, *Taare Zameen Par*, about a young boy struggling to learn in school, brought the term "dyslexia" into the mainstream, seeking to raise awareness, clarify some misconceptions, and reduce the stigma associated with it ("Pain of Dyslexia," 2008). Asked if she thought *Taare Zameen Par* had had any effect, Kate Currawala, the founder and director of the Maharashtra Dyslexia Association (MDA), replied,

> I think it raised awareness and created empathy. It made a difference in attitudes. People expected the condition to be more visible and many people didn't connect it with behavior problems. But I don't think it reduced the stigma.

Two, sometimes children from non-English-speaking homes, many of whom may also be poor or *Dalit*, may get labeled because they perform at a lower level than students who come to school knowing English, not because they have a disability (Mukhopadhyay & Sriprakash, 2011).

Mukhopadhyay and Sriprakash (2011) found that when government schools failed to meet standards of school effectiveness, teachers began assessing increasing proportions of students from marginalized groups as "failing." Another study found that the RTE Act mandate to reserve 25% seats in private schools toward ensuring educational access for children from economically and socially marginalized communities resulted in many children being subjected to overt discrimination by teachers who viewed them as "slow learners," "weak" or "unteachable" (CES, 2014). These children became reluctant to seek clarifications because they were "scared that teachers would scold, beat or insult them, or that peers would make fun of them for what they did not know" (CES, 2014, p. 56).

Three, many schools do not offer in-school services, and most special schools charge high fees making them unaffordable to most children; so students get labeled but no support (CES, 2014). In the US, programs and practices for LD students emerge from a resource-rich model of service provision that is often incompatible with the Indian realities. For instance, schools in India are expected to provide certain modifications required by law, but the process for procuring these services can be complicated and discouraging for most families, especially those from low-income backgrounds (Ghai, 2006). The shortage of LD specialists means that most schools do not provide on-site support services, and many remedial special schools charge high fees for the specialized services, making them unaffordable to low-income students (CES, 2014). Further, the number of officially recognized languages in India makes creating a standardized assessment measure for the specific detection and educational intervention of children with LD problematic (Narayan et al., 2003; Ramaa, 2000; Unni, 2012). Till recently, the American publishing company, Pearson, now based in India, was marketing diagnostic tools normed on the 2000 US Census to assess dyslexic students in Indian schools ("Now, a toolkit," 2012), suggesting that Indian children were being assessed and diagnosed as LD because they failed a test normed on US standards.

Four, despite local research in the LD field emphasizing the importance of recognizing the overlap of language with learning disability within India's multilingual context, national policies and legislation for children with disabilities do not acknowledge this imbrication. For instance, as a signatory to international declarations that "emphasize the need for fundamental educational policy shifts to enable general schools to include children with disabilities" (Ministry of Human Resource Development,

2019, p. 60), India's EFA policy stipulates that all educational institutions recognized or funded by the government provide an inclusive education to children with disabilities. To this end, schools are required to "detect specific learning disabilities in children at the earliest and take suitable pedagogical and other measures to overcome them" (p. 60). However, the focus on multilingual education, mother tongue education and bridge courses is for tribal children only, and the educational provision in the most appropriate languages and modes and means of communication is only for students who are blind or deaf or both. The category of speech and language disability refers to a permanent disability arising from conditions like laryngectomy or aphasia, affecting components of speech and language due to organic or neurological causes.

Just as research indicated classist and racist elements within the phenomenon of labeling of learning disabilities in the US, there appears to be classist/casteist and linguistic elements to the phenomenon in India. Who are these children? What benefits do they accrue from this label? Why is the label itself based on the American definition? Why does the emergence of this category look similar to that in the US? Are services differentiated by class/caste or language difference, as in the US? These are some questions this study attempted to answer.

The Methodology

Disability scholars have noted the separation of disability and education spaces in India till recently (Chennat, 2019c; Naraian, 2019), largely because children with disabilities were the last entrants to schools (UNESCO, 2019; Vaidya, 2020) and interpretations of inclusion involved other marginalized student populations as well (Singal, 2006). This recent research has focused on the inclusion of children with disabilities in general education inclusive settings, examining, for instance, teacher preparedness and attitudes (Ahsan et al., 2013; Limaye, 2019; for a review, see Singal, in press). Separate research has focused on the exclusion of other marginalized communities in general education, often resulting from teacher biases (e.g., CES, 2014; Mukhopadhyay & Sriprakash, 2011; Nambissan, 2009). However, there is little research that explores the intersectionality of socio-economic status with disability in inclusive general education settings, and the effects of teacher attitudes or biases on marginalized students who are struggling academically and may end up being labeled "learning disabled." The postcolonial legacy of the English language and the issue of medium of instruction, as a further complication for students from non-English-speaking backgrounds, is also under-researched (Boruah, 2017; Shenoy & Pearson, 2018).

The study sought to explore if English as a medium of instruction was a confounding factor in the learning and labeling of children, particularly for children from low socio-economic backgrounds who might be less conversant with English, and to determine the availability of support services. It used a case study approach (Schwandt & Gates, 2018; Yin, 2014), in which a unit of analysis at the micro-level, such as a school, is studied to understand a sociocultural phenomenon occurring at a macro-level, and applied qualitative research methods of participant observations, interviews and document analysis (Charmaz, 2014; Glesne, 2016).

The Setting

The study was conducted in Mumbai, in the western state of Maharashtra, over a four-month period. The Maharashtra government has been in the front of the LD field. It recognized LD as a disability category in 1996, was the first in India to formally grant students with LD the benefit of instructional modifications (Karande et al., 2011), and, in 2018, the Bombay High Court mandated screening students for LD before Standard 3. These policies suggested a robust system of supports.

The study focused primarily on three low-fee-paying (LFP) English-medium schools, with three high-tier English-medium schools[5] being included as points of comparison. Five of the schools, including all three LFP schools, were located within a two-kilometer radius of each other in south central Mumbai[6] that has seen extremely rapid development since the 1990s. The LFP schools were in the poorest section of the neighborhood, the *basti*, while two high-tier schools were in the more middle-class parts. The third high-tier school was located in more elite South Mumbai, about 20 km away.

Mumbai is India's largest city, and also the densest, with over 18 million people, according to the 2011 national Census, within an area of 603.4 square kilometers (233 sq. miles) (Limaye, 2019). Newly independent India's decision to follow the path of urbanization toward development encouraged by international development agencies precipitated large-scale rural to urban migration (Bhagat, 2017; Juneja, 2017).[7] However, few urban planning policies were put into place to accommodate this onrush of people to the outskirts of large metropolitan cities, leading to the creation of slums or *bastis* that grew larger, without basic amenities of water, sewerage, or electricity (Bhagat, 2017; Bhatia, 2018; Garrett, 2000).[8] According to the 2011 census, Mumbai's slum population is estimated to be 41.3% of all households (Juneja, 2017).

The *basti* in which the study took place developed in much the same way as other slums in Mumbai. Originally a predominantly Hindu village, it had become the resettlement point for many Muslim families fleeing communal riots in the city in 1992, changing its demographic profile. Later, migrant families from other parts of Maharashtra and other states, such as Uttar Pradesh in the north and Tamil Nadu in the South, had also settled here. By 2017, when this study was undertaken, the original village had grown into a bustling community of low-income housing, from *chaal* or chummery to apartments atop small retail shops[9] which lined the streets. The streets were always crowded during the day with pedestrians, autorickshaws, cars and buses jostling for space, avoiding potholes and each other. During the monsoons, lack of drainage would cause rainwater to collect in the streets, combining with mud and refuse to become puddles of dark, viscous water that children waded happily through on their way back from school. Behind one of the schools, on the other side of an open sewer were the *jhuggis* or hutments of people living in the *basti*, and in the far background the skyscraper homes of the middle class and affluent.

Data Collection

The study was conducted in India between 2017 and 2018 and received Institutional Research Board (IRB) approval and Government of India permission. Since the study intended to explore the influence of learning in English on low-income students who came from linguistically diverse backgrounds, the primary focus was on three LFP, private, English-medium schools through a case study approach, which provides opportunities for an in-depth exploration of a unit of analysis at the micro level to understand a sociocultural phenomenon occurring at a macro-level (Schwandt & Gates, 2018; Yin, 2014). The three schools, which expressed willingness to participate in the study, were selected as being representative of LFP, private, English-medium schools that served students from socio-economically disadvantaged and linguistically diverse backgrounds. Following case study or replication logic, the schools were added sequentially until reaching data saturation (Schwandt & Gates,

2018; Small, 2009). As the study developed in the inductive process of qualitative research (Creswell, 2007; Corbin & Strauss, 2015), three upper-tier English-medium schools, one of which was a special school for students with learning disabilities, were included to a lesser extent to provide points of comparison, as were supplementary services, such as tuition centers and private tuition classes, which comprise a "shadow education" (Sarangapani & Winch, 2010), and diagnostic clinics.

Qualitative research methods of extensive observations and in-depth interviews were used (Charmaz, 2014; Corbin & Strauss, 2015; Glesne, 2016), since the purpose was to understand the "social contexts in which policy and reform imperatives are reshaped in schools, often in unintended or unexpected ways" (Sriprakash, 2011, p. 6). Further, the study foregrounded teachers at the LFP schools as research on teachers' perspectives of educational change in the Indian context, particularly of women teachers at LFP schools, is limited (Manjrekar, 2013; Sriprakash, 2011). Classroom observations were followed up with individual interviews with the teachers in an iterative process to learn how they interpreted their own instruction and implementation of inclusion (Mukhopadhyay, 2020; Singal, in press).

Classroom observations were conducted in grades 3 to 5 in the LFP schools, at the administrators' suggestion, since they felt students would be most likely to demonstrate academic difficulties by then. I began by observing in three sections of each grade level, and then focusing on three teachers in each school who expressed willingness to participate in the study. At this point, the teachers were asked to identify students who they thought were struggling academically. The teachers were interviewed before and/or after the observation and asked about their instruction and their assessment of these students' performance, as well as their demographic details, including level of education. I also spoke to teachers at other grade levels in the teachers' lounge and observed some remedial class sessions conducted in the schools.

Interviews (Brinkmann, 2015) were conducted with a variety of participants. The teachers in each of the classrooms I observed were interviewed. This included general education teachers and, in the upper-tier schools, the special education teachers as well. In addition, interviews were conducted with principals at the schools, tutors at the tuition centers and private tuition classes, and parents, clinicians at three diagnostic clinics or assessment centers, professionals about the Dyslexia Assessment for

Languages of India (DALI) and representatives at local and national agencies, such as the Maharashtra Dyslexia Association (MDA) and Teach For India and Pratham, which work with LFP schools. Jot notes were maintained during the interviews, taking care to keep as close as possible to the original words and phrases used by the participants (Brinkmann, 2015).

The parents were contacted through the teachers at the LFP schools. If the parents agreed to speak to me, we would move to a corner of the classroom or step outside the school and conduct the interview there. Sometimes other parents would join in as I interviewed a parent. I used my own connections and the special school to identify and interview upper-class parents. Interviews were conducted in Hindi[10] with the LFP parents and in English with all other participants. I spoke to university faculty in education or disability studies, sometimes as member checks on what I was learning. In the tradition of qualitative research, which considers all data to be significant (Bogdan & Biklen, 2006; Creswell, 2007), I also documented informal interactions and conversations. For instance, I asked visitors at the university guesthouse I was at, who were attending weekend courses, conferences and other university-related events, about their thoughts on the Indian educational system and their own experiences with schooling.[11]

Document analysis (Bowen, 2009) was conducted with national policies, student work samples to identify patterns in linguistic and conceptual errors, and class textbooks to determine vocabulary levels and curricular content. I assessed students' reading skills and comprehension informally as they copied notes from the blackboard, asking them to read what they'd written and then explain to me what they had read in a language they felt more comfortable using, whether English or Hindi. When I found students struggling, I explained or reviewed the concept in Hindi to assess if language was a barrier to students' comprehension. The participant observations, semi-structured interviews, informal conversations and document analysis provided depth and detail (Patton, 1980) and allowed for the triangulation of data (Denzin, 2012).

Data Analysis

Jot notes taken during interviews and observations were written up more extensively as fieldnotes totaling close to 400 single-spaced, typewritten pages. Data was analyzed using a recursive, constant comparison approach (Charmaz, 2014; Corbin & Strauss, 2015; Glesne, 2016), punctuated with analytic memos to identify themes and when data saturation had occurred. For instance, as I began to analyze the data I was collecting at the LFP schools, I found that English as a medium of instruction was a significant factor in creating students' academic difficulties in these schools. When assessing them informally, I noticed that they understood the concepts much more easily in Hindi. In sharing these findings with university faculty, as part of member check processes, I was advised to include some upper-tier schools where students' fluency in English might be less of an impacting factor. This led me to spend some time observing at three upper-tier schools.

A combination of manual coding and NVivo, a qualitative data analysis software package, was used to develop codes and categories for a thematic analysis toward a grounded theory (Charmaz, 2014; Creswell, 2007; Kolb, 2012). Both deductive coding, or top-down codes based on a theoretical framework, and inductive coding, or bottom-up development of codes (Charmaz, 2014; Saldaña, 2021), were applied toward identifying themes that ran across the data. For instance, some initial codes included student descriptors, such as "coming from slum areas," "uneducated parents" and "ten rupee students," which were categorized as differences in caste/class status, developing into a theme of the school as a reflection of societal hierarchy and teachers' limited agency within recent neoliberal education reforms. Codes such as "choral recitation," "copying notes" and "paper pattern reviews" were categorized as teachers' responses to students' difficulties with learning English simultaneously with learning in English, generating a theme on the overlap of language and learning. These themes, among others, led to a grounded theory of teachers' perceptions of and efforts to teach first-generation students in English-medium LFP schools within the context of education for all.

Structure of the Book

As Baglieri and Shapiro (2017) commented, if there is no societal expectation that everyone be literate, children would not be labeled as unable to read or learning disabled. Universal access to education and inclusive education thus becomes a double-edged sword. This is not to suggest that education should be elitist and not be available to all, but to caution against adopting such an umbrella policy framework without considering its unexpected consequences and providing for them. In the following chapters, this book explores some of the nuances of these unexpected and negative consequences, presenting an analysis of the interplay of language and caste/class factors for poor and lower-caste students in LFP, private English-medium schools and identifying how outcomes are different for middle-class and elite students at upper-tier schools. Using a sociocultural approach of nested macro, meso and micro systems at the global, national and school levels (Bourdieu & Passeron, 1990), this analysis is placed within the larger contexts of global conditions and national milieus that explain the impetus for low-income parents to seek private, English-medium schools.

Chapter 2 provides the global backdrop against which many policies, programs and principles have been transferred from the global North and implemented in the global South, and India in particular. It argues that a universal template of development which assigns a linear progression places countries that fall toward the lower half of this continuum at a disadvantage and under the tremendous stress of playing catch-up. Further, educational systems that are proving problematic within their countries of origin are being transplanted uncritically to meet these development goals, resulting in many of the same problems being replicated and new problems emerging in these new contexts. Chapter 3 examines these problems within the national context, analyzing the government's EFA programs and language education policies, such as the Three-Language Formula policy, and legislations like the 2009 Right to Education Act and the 2016 Rights of Persons with Disabilities Act, which included provisions for students with learning disabilities for the first time. It argues that the western-centric stance embedded in these policies fails to consider cultural relevance and individual student variability.

The next two chapters examine the implications of this global and national framework at the micro-level of the school. Chapter 4 draws parallels with the societal hierarchies, both traditional and more recent as India transitions toward development, and the hierarchies in schools. It presents teachers' perceptions of their students and their families as coming from "slum areas" as a reflection of these larger social factors of caste/class and even religious difference, and teachers' own limited agency as women within an increasingly gendered, low-paid profession. Chapter 5 explores how English as a medium of instruction adversely affects the quality of education students in LFP schools receive, in contexts where teachers have limited pedagogical training in bilingual education, and rigid curriculum and large class sizes inexorably steer them toward "teaching-in-translation" and "teaching to the test" (Bhattacharya, 2013; Sriprakash, 2011). Although students' academic struggles emerge from trying to learn in English and learn English simultaneously, their failure is perceived as cognitive deficits and they are relegated to the back of the class as "back-benchers" and "slow learners."

Chapter 6 analyzes the complications arising from an identification and assessment process that is steeped in international norms and western-centric epistemological assumptions about the universality of knowledge. Instead of receiving supports at primary grade levels, the learning disability certificate has been "domesticated" (Gupta, 2014) to get mandated supports during the more crucial and fiercely competitive secondary school-leaving exam. Chapter 7 draws specific conclusions to present alternatives to the current challenges and dilemmas. It suggests possibilities for recognizing and anticipating that placing new expectations on children may lead to some children being unable to meet these expectations and putting into place humanistic modifications to support them instead of allowing additional marginalization and stigmatization to occur.

Conclusion

Applying the lenses of postcolonialism and disability studies in a study of educational systems for children with disabilities, we can begin to identify the influence of the larger contexts of globalization and international development, and the more recent neoliberal economic reforms on the creation of the category of learning disabilities and its importation to India, on the assessment process and the measurement tools being used, and the differentiated benefits based on socio-economic factors such as caste/class or language, in terms of access to English and services.

The intent is not to deny the possibility that, with the new expectation that all children learn to read, write and do math—and, perhaps, learn all this in English, some children may struggle to acquire these skills. Indeed, this study found students who did. However, locating within the child what we, as a society, have now deemed as an inadequacy and blaming the child for a so-called deficit is neither helpful to the child nor society. Instead, we need to explore how the current context of globalization has brought us to a point where this expectation is being placed on all children, regardless of the value of these skills. We need to analyze the consequences of the tyranny of the norm, of cookie-cutter educational systems. We need to examine the implications of putting structures into place for students who struggle to learn, for reasons that have little to do with cognitive ability, which are causing a new form of exclusion even as we espouse inclusion.

Notes

1. This term was not used by any participant in this study, except once when the director used it in a formal address to parents in Hindi during a celebratory ceremony at a learning disability child guidance center.
2. Although at 7%, learning disability is a low incidence disability in India, whereas in the US, it is a high incidence disability constituting over 50% of all children with disabilities, the unquestioning westward stance, sometimes results in SLD being referred to as "one of the high incidence disabilities" in India as well (Panshikar, 2019, p. 175).
3. Prevalence rates for learning disabilities differ according to the source. Panshikar (2019), for instance, notes that "various studies done have documented the prevalence of SLD between 3 and 14% of school-going population (Bandla et al., 2017; Mogasale & Mogasale, 2012; Padhya et al., 2016)" (p. 176).
4. Samagra Shiksha Abhiyan subsumes the three existing schemes of Sarva Shiksha Abhiyan (SSA), Rashtriya Madhyamik Shiksha Abhiyan (RMSA) and Teacher Education. SSA focuses on children at elementary level (Grades 1–8) and RMSA on children at secondary level (Grades 9–12), while the Teacher Education component oversees preservice and in-service training and academic resource support to elementary and secondary school teachers and schools (UNESCO, 2019).
5. Schools are tiered primarily on the basis of the difference in fee structure: high-tier schools charge higher fees, affordable only to students from middle and upper class backgrounds, while low-fee-paying schools cater largely to students from low socio-economic backgrounds (Shenoy & Wagner, 2019).
6. In the context of development in developing countries, it is important to note that, while this area is deemed a suburb of Mumbai, the term "suburb" has very different connotations from that in the

US. Whereas in the US the suburb is a highly desirable location which most families aspire to flee to as soon as they are economically able, in India, the suburbs are the neglected outskirts of cities, where migrant labor from the rural areas were forced to settle, when urbanization first began. Over time, as the city grew, sections of these suburbs increased in real estate value and social status and began to incorporate middle-class and affluent community settlement (Juneja, 2017).

7. Juneja (2017) notes that 25% of the entire Indian urban population is concentrated in just five "mega" cities—Mumbai, Delhi, Kolkata, Chennai, and Bengaluru, and that, over the years, internal urban growth has increased more than external growth through migration.
8. State government policies, such as the Maharashtra Slum Improvement Act of 1971 and the Slum Rehabilitation Program, have attempted to mitigate some of these inequities over the years (Bhatia, 2018). Yet, many *basti* residents are not granted land ownership rights, despite having lived there for several decades, making life in slums uncertain and precarious (Bhatia, 2018; Boo, 2012).
9. Most of the shops sold Western clothes, including inexpensive, knock-off denim pants and synthetic shirts and blouses. Hairdresser shops displayed pictures of has-been Hollywood stars, like Alain Delon, and current Bollywood movie stars, like Shah Rukh Khan. Perhaps because of the large Muslim population, there were several butcher's shops, some of which sold prepared food, like *kebabs* and *biryani*. A few of these shops had a goat tethered in the front to indicate that the meat sold was mutton and not beef, a response to the chilling lynchings orchestrated by right-wing Hindutva party members against Muslims and *Dalits* in a campaign to protect the sacred Hindu cow.
10. Technically the interviews were conducted in Hindustani, which is a hybrid of the more formal languages of Hindi and Urdu, and in Mumbai, includes a smattering of Marathi as well; it is the language most commonly spoken in every day informal interactions (Mohan, 2017).
11. Many spoke of the difficulties, even trauma, they had faced when they transitioned from regional language education in secondary school to higher education in the English language, but averred

that this had helped them to become equally fluent and truly bilingual in both. Others talked about the advantage of being fluent in reading and speaking in at least two languages, keeping English to the academic arena. Some spoke about losing their mother tongue and acquiring only a rudimentary level of fluency in the regional language, as they became proficient in English.

References

10% of kids in India have learning disability: Experts say. (2012, January 27). *Times of India*. http://articles.timesofindia.indiatimes.com/2012-01-27/chennai/30669918_1_inclusive-education-disability-autism

Addlakha, R. (Ed.). (2013). *Disability studies in India: Global discourses, local realities*. Routledge.

Ahsan, M. T., Deppeler, J. M., & Sharma, U. (2013). Predicting pre-service teachers' preparedness for inclusive education: Bangladeshi pre-service teachers' attitudes and perceived teaching-efficacy for inclusive education. *Cambridge Journal of Education, 43*(4), 517–535. https://doi.org/10.1080/0305764X.2013.834036

Alur, M. (2002). "They did not figure": Policy exclusion of disabled people in India. *International Journal of Inclusive Education, 6*, 101–112.

Anand, S. (2015). Corporeality and culture: Theorizing difference in the South Asian context. In S. Rao & M. Kalyanpur (Eds.), *South Asia and Disability Studies: Redefining boundaries and extending horizons* (pp. 158–174). Peter Lang.

Annamma, S. A., Ferri, B. A., & Connor, D. J. (2018). Disability critical race theory: Exploring the intersectional lineage, emergence, and potential futures of discrit in education. *Review of Research in Education, 42*, 46–71. https://doi.org/10.3102/0091732X18759041

ASER. (2019). *Annual status of education report, 2018*. Pratham, Anthem Press.

Arne, E., & Ingstad, B. (Eds.). (2011). *Disability and poverty: A global challenge*. Policy Press.

Baglieri, S., & Shapiro, A. (2017). *Disability Studies and the inclusive classroom: Critical practices for embracing diversity in education* (2nd ed.). Routledge.

Bhagat, R. B. (2017). *Migration and urban transition in India: Implications for development*. United Nations Secretariat. https://www.un.org/en/development/desa/population/events/pdf/expert/27/papers/V/paper-Bhagat-final.pdf

Bhatia, S. (2018). *Decolonizing Psychology: Globalization, Social Justice, and Indian Youth Identities*. Oxford University Press.

Bhattacharya, U. (2013). Mediating inequalities: Exploring English-medium instruction in a suburban Indian village school. *Current Issues in Language Planning, 14*(1), 164–184. https://doi.org/10.1080/14664208.2.

Blanchett, W. J., Klingner, J. K., & Harry, B. (2009). The intersection of race, culture, language, and disability: Implications for urban education. *Urban Education, 44*(4), 389–409. https://doi.org/10.1177/0042085909338686

Bogdan, R., & Biklen, S. K. (2006). *Qualitative research for education: An introduction to theories and methods* (5th ed.). Pearson.

Boo, K. (2012). *Behind the beautiful forevers: Life, death and hope in a Mumbai undercity*. Random House.

Boruah, P. (2017). Learning English in a low-cost semi-urban English-medium school in India: Challenges, interaction patterns and domains of use. In H. Coleman (Ed.), *Multilingualisms and Development* (pp. 289–306). British Council.

Bourdieu, P., & Passeron, J. C. (1990). *Reproduction in education, society and culture*. Sage.

Bowen, G. A. (2009). Document analysis as a qualitative research method. *Qualitative Research Journal, 9*(2), 27–40.

Bowles, S., & Gintis, H. (2011). *Schooling In capitalist America: Educational reform and the contradictions of economic life*. Haymarket Books.

Brinkmann. (2015). *The interview. The SAGE handbook of qualitative research*. Sage.

Center for Equity Studies (CES). (2014). *India exclusion report, 2013–14*. Books for Change.

Charmaz, K. (2014). *Constructing grounded theory* (2nd ed.). Sage.

Chennat, S. (2019a). Disability studies: The context. In S. Chennat (Ed.), *Disability inclusion and inclusive education* (pp. 1–19). Springer. https://doi.org/10.1007/978-981-15-0524-9_1

Chennat, S. (2019b). Sociology of disability. In S. Chennat (Ed.), *Disability inclusion and inclusive education* (pp. 21–37). Springer. https://doi.org/10.1007/978-981-15-0524-9_2

Chennat, S. (2019c). Conceptualizing disability inclusion. In S. Chennat (Ed.), *Disability inclusion and inclusive education* (pp. 39–62). Springer. https://doi.org/10.1007/978-981-15-0524-9_3

Collin, R., & Apple, M. (2010). Globalization, schooling, and literacies in an unsettled age. In S. Tozer, B. P. Gallegos, A. Henry, M. Bushnell Greiner, & P. Groves Price (Eds.), *Handbook of research in the social foundations of education* (pp. 296–307). Routledge.

Connor, D. J., Valle, J. W., & Hale, C. (2015). *Practicing disability studies in education: Acting towards social change*. Peter Lang.

Corbin, J., & Strauss, A. (2015). *Basics of qualitative research: Techniques and procedures for developing grounded theory* (4th ed.). Sage.

Creswell, J. W. (2007). *Qualitative inquiry and research design: Choosing among five traditions*. Sage.

Das, J. P. (2008). Foreword. In K. Thapa & G. M. van der Aalsvoort (Eds.), *Perspectives on learning disabilities in India: Current practices and prospects* (pp. ii–v). Sage.

Denzin, N. K. (2012). Triangulation 2.0. *Journal of Mixed Methods Research, 6*, 80–88. https://doi.org/10.1177/1558689812437186

Dudley-Marling, C. (2004). The social construction of learning disabilities. *Journal of Learning Disabilities, 37*(6), 482–489.

Duhaney, L. M. G. (2014). Disproportionate representation in special education: A persistent stain on the field. In F. E. Obiakor & A. F. Rotatori (Eds.), *Contemporary perspectives in special education: Multicultural education for learners with special needs in the 21st century* (pp. 15–40). Information Age Publishing.

Erevelles, N. (2011). *Disability and difference in global contexts: Enabling a transformative body politic*. Palgrave Macmillan.

Ferri, B. A., & Connor, D. J. (2006). Tools of exclusion: Race, disability and (re)segregated education. *Teachers College Record, 107*(3), 453–474.

Foucault, M. (1972). *The archaeology of knowledge and the discourse on language*. Pantheon Books.

Gabel, S. (2004). South Asian Indian cultural orientations towards mental retardation. *Mental Retardation, 42*(1), 12–25.

Garrett, L. (2000). Filth and decay: Pneumonic plague hits India and the world ill responds. In L. Garrett (Ed.), *Betrayal of trust: The collapse of global public health* (pp. 15–48). Hyperion.

Ghai, A, (2006). Education in a globalising era: Implications for disabled girls. *Social Change, 36*(3), 161–176. https://doi.org/10.1177/004908570603600306

Ghai, A. (2020). Ignorance of disability: Some epistemological questions. In N. Mehrotra (Ed.), *Disability Studies in India* (pp. 75–91). Springer. https://doi.org/10.1007/978-981-15-2616-9_1

Giroux, H. A. (1983). *Theory and resistance in education: A pedagogy for the opposition*. Bergin.

Glesne, C. (2016). *Becoming qualitative researchers: An introduction* (5th ed.). Pearson.

Goddess of English. (2011, February 15). *An 'English goddess' for India's down-trodden*. https://www.bbc.com/news/world-south-asia-12355740.

Gould, S. J. (1996). *The mismeasure of man*. W.W. Norton.

Grech, S. (2009). Disability, poverty and development: Critical reflections on the majority world debate. *Disability & Society, 24*, 771–784.

Grosfoguel, R. (2013). The structure of knowledge in westernized universities epistemic racism/sexism and the four genocides/epistemicides of the long

16th Century. *Human Architecture: Journal of the Sociology of Self-Knowledge, 11*(1), 73–90.

Gupta, R. (2014). Change in teaching practices: Case of phonics instruction in India. *Procedia: Social and Behavioral Sciences, 116*, 3911–3915. https://doi.org/10.1016/j.sbspro.2014.01.865

Hammad, T., & Singal, N. (2015). Disability, gender, and education: Exploring the impact of education on the lives of women with disabilities in Pakistan. In S. Rao & M. Kalyanpur (Eds.), *South Asia and Disability Studies: Redefining boundaries and extending horizons* (pp. 197–228). Peter Lang.

Hardt, M., & Negri, A. (2000). *Empire*. Harvard University Press.

Harry, B. (2014). The disproportionate placement of ethnic minorities in special education. In L. Florian (Ed.), *The SAGE Handbook of special education* (Vol. 1, 2nd ed., pp. 73–95). Sage.

Illich, I. (1971). *Deschooling society*. Marion Boyars.

Jackson, R. F., & Lee, C. M. (1992). *Faking it: A look into the mind of a creative learner*. Heinemann.

Jangira, N. K. (2002). Special educational needs of children and young adults: An unfinished agenda. In M. Alur & S. Hegarty (Eds.), *Education and children with special needs: From segregation to inclusion* (pp. 67–76). Sage.

Jha, M. M. (2004). Inclusive education and the common school in India. In C. S. Mohapatra (Ed.), *Disability management in India: Challenges and commitments* (pp. 160–171). National Institute for the Mentally Handicapped.

Juneja, N. (2017). Education in urban areas. In K. Kumar (Ed.), *Routledge handbook of education in India, debates, practices, and policies* (pp. 26–42). Routledge. https://www.routledgehandbooks.com/doi/10.4324/9781315107929.ch2

Kalia, V., & Reese, E. (2009). Relations between Indian children's home literacy environment and their English oral language and literacy skills. *Scientific Studies of Reading, 13*(2), 122–145. https://doi.org/10.1080/10888430902769517

Kalyanpur, M. (1996). The influence of western special education on community-based services in India. *Disability & Society, 11*(2), 249–270.

Kalyanpur, M. (2008a). The paradox of majority under-representation in special education in India: Constructions of difference in a developing country. *Journal of Special Education, 42*(1), 55–64.

Kalyanpur, M. (2008b). Equality, quality and quantity: Challenges in inclusive education policy and service provision in India. *International Journal of Inclusive Education, 12*(3), 243–263.

Karande, S., Sholapurwala, R., & Kulkarni, M. (2011). Managing specific learning disability in schools in India. *Indian Pediatrics, 48*, 517–520.

Karanth, P. (2008). Learning disability and language learning. In K. Thapa, G. M. van der Aalsvoort, & J. Pandey (Eds.), *Perspectives on learning disabilities in India: Current practices and prospects* (pp. 80–96). Sage.
Karna, G. N. (2001). *Disability studies in India: Retrospects and prospects*. Gyan Publishing.
Kliewer, C., & Fitzgerald, L. M. (2001). Disability, schooling, and the artifacts of colonialism. *Teachers College Record, 103*(3), 450–470.
Kolb, S. M. (2012). Grounded theory and the constant comparative method: Valid research strategies for educators. *Journal of Emerging Trends in Educational Research and Policy Studies (JETERAPS), 3*(1), 83–86.
Kozleski, E. B. (2020). Disrupting what passes as inclusive education: Predicating educational equity on schools designed for all. *The Educational Forum, 84*(4), 340–355. https://doi.org/10.1080/00131725.2020.1801047
Ladwig, J. G., & McPherson, A. (2017). The anatomy of ability. *Curriculum Inquiry, 47*(4), 344–362. https://doi.org/10.1080/03626784.2017.1368352
Limaye, S. (2019). Unheard voices: Schooling experiences of parents having children with disabilities in India. In M. J. Schuelka, C. J. Johnstone, G. Thomas, & A. J. Artiles (Eds.), *The SAGE Handbook of diversity and inclusion in education* (pp. 557–570). Sage.
Manjrekar, N. (2013). Women school teachers in new times: Some preliminary reflections. *Indian Journal of Gender Studies, 20*(2), 335–356.
Maudslay, L. (2014). Inclusive education in Nepal: Assumptions and reality. *Childhood, 21*(3), 418–424. https://doi.org/10.1177/0907568213514778
McDermott, R., Edgar, B., & Scarloss, B. (2011). Global norming. In A. Artiles, E. Kozleski, & F. Waitoller (Eds.), *Inclusive Education in five continents* (pp. 223–235). Harvard Education Press.
Mehrotra, N. (Ed.). (2020a). *Disability Studies in India*. Springer. https://doi.org/10.1007/978-981-15-2616-9_1
Mehrotra, N. (2020b). Situating disability studies: A prolegomenon. In N. Mehrotra (Ed.), *Disability Studies in India* (pp. 1–22). Springer. https://doi.org/10.1007/978-981-15-2616-9_1
Mitra, S. (2006). The capability approach and disability. *Journal of Disability Policy Studies, 16*(4), 236–247.
Ministry of Human Resource Development. (2019). *Draft National Education Policy 2019*. Government of India.
Mooney, J. (2007). *The short bus: A journey beyond normal*. Henry Holt.
Motha, S. (2014). *Race, empire, and English language teaching: Creating responsible and ethical anti-racist practice*. Teachers College Press.
Mohan, P. (2017). The road to English: The slow migration of the EWS child to elite India. In M. J. Tiwary, S. Kumar, & A. K. Mitra (Eds.), *Dynamics of*

inclusive classrooms: Social diversity, inequality and school education in India (pp. 203–215). Orient BlackSwan.

Mukherjee, S. (2016). *The gene: An intimate history*. Scribner.

Mukhopadhyay, L. (2020). Translanguaging in primary level ESL classroom in India: An exploratory study. *International Journal of English Language Teaching, 7*, 2. https://doi.org/10.5430/ijelt.v7n2p1

Mukhopadhyay, R., & Sriprakash, A. (2011). Global frameworks, local contingencies: Policy translations and education development in India. *Compare: A Journal of Comparative and International Education, 41*(3), 311–326. https://doi.org/10.1080/03057925.2010.534668, http://www.informaworld.com.

Nambissan, G. B. (2009). *Exclusion and discrimination in schools: Experiences of Dalit children*. Indian Institute of Dalit Studies and UNICEF.

Naraian, S. (2016). Spatializing student learning to reimagine the "place" of inclusion. *Teachers College Record, 118*(3), 1–46.

Naraian, S. (2019). Diversifying theoretical commitments for a transnational inclusive education: Lessons from India. In N. Singal, S. T. Johansson & P. Lynch (Eds.), *Education and disability in the global South: New perspectives from Africa and Asia* (pp. 107–126). Bloomsbury Academic.

Narayan, J., Thressiakutty, A. T., Haripriya, C., Reddy, K. G., & Sen, N. (2003). *Educating children with learning problems in primary schools: Resource book for teachers*. National Institute for the Mentally Handicapped.

National Joint Committee on Learning Disabilities (NJCLD). (2016). *What are LD? National Joint Committee on Learning Disabilities Definition of Learning Disabilities*. NJCLD Online. https://njcld.org/ld-topics/

Now, a toolkit for dyslexic students in Indian schools. (2012, July 25). *The Times of India*. http://articles.timesofindia.indiatimes.com/2012-07-25/news/32847749_1_tool-kit-dyslexic-english-medium-schools

The Pain of Dyslexia, as Told by Bollywood. (2008, June 4). *Washington Post*. http://www.washingtonpost.com/wp-dyn/content/article/2008/06/03/AR2008060303201.html

Panshikar, A. (2019). Specific learning disability: A hidden disability. In S. Chennat (Ed.), *Disability inclusion and inclusive education* (pp. 175–193). Springer. https://doi.org/10.1007/978-981-15-0524-9_9

Pandey, J., & Pant, S. K. (2008). The marginalized Dalits: Disadvantages in learning. In K. Thapa & G. M. van der Aalsvoort (Eds.), *Perspectives on learning disabilities in India: Current practices and prospects*. Sage.

Parekh, G. (2017). The tyranny of "ability." *Curriculum Inquiry, 47*(4), 337–343. https://doi.org/10.1080/03626784.2017.1383755

Patton, M. Q. (1980). *Qualitative evaluation methods*. Sage.

Ramaa, S. (2000). Two decades of research on learning disabilities in India. *Dyslexia, 6*, 268–283.

Said, E. W. (1979). *Orientalism*. Vintage Books.
Saldaña, J. (2021). *The coding manual for qualitative researchers* (4th ed.). Sage.
Sarangapani, P. M., & Winch, C. (2010). Tooley, Dixon and Gomathi on private education in Hyderabad: A reply. *Oxford Review of Education, 36*(4), 499–515. https://doi.org/0.1080/03054985.2010.495465
Schwandt, T. A., & Gates, E. F. (2018). Case study methodology. In N. K. Denzin & Y. S. Lincoln (Eds.), *The SAGE handbook of qualitative research* (5th ed., pp. 341–358). Sage.
Shenoy, S., & Wagner, R. K. (2019). Language and literacy practices that influence bilingual and bi-literate acquisition in L1 Kannada and L2 English in Bangalore, India. In R. M. Joshi, & C. McBride (Eds.), *Handbook of Literacy in Akshara Orthography, Literacy Studies, 17*. https://doi.org/10.1007/978-3-030-05977-4_19
Shenoy, S., & Pearson, D. (2018). School culture and its impact on special education practices in Bangalore. *The Journal of the International Association of Special Education, 18*(1), 49–63.
Simpson, M. K. (2007). From savage to citizen: Education, colonialism and idiocy. *British Journal of Sociology of Education, 28*(5), 561–574.
Singal, N. (2005). Mapping the field of inclusive education: A review of the Indian literature. *InTernational Journal of Inclusive Education, 9*, 331–350.
Singal, N. (2006). Inclusive education in India: International concept, national interpretation. *International Journal of Disability, Development and Education, 53*, 351–369.
Small, M. L. (2009). "How many cases do I need?": On science and the logic of case selection in fieldbased research. *Ethnography, 10*, 5–38.
Sriprakash, A. (2011). Being a teacher in contexts of change: Education reform and the repositioning of teachers' work in India. *Contemporary Education Dialogue, 8*(1), 5–31.
Staples, J. (2020). Decolonising disability studies? Developing South Asia-specific approaches to understanding disability. In N. Mehrotra (Ed.), *Disability Studies in India* (pp. 25–41). Springer. https://doi.org/10.1007/978-981-15-2616-9_1
Terzi, L. (2010). *Justice and equality in education: A capability perspective on disability and special educational needs*. Continuum.
Thapa, K. (2008). Learning disabilities: Issues and concerns. In K. Thapa, G. M. van der Aalsvoort, & J. Pandey (Eds.), *Perspectives on learning disabilities in India: Current practices and prospects* (pp. 23–47). Sage.
Thapa, K., Aalsvoort, G.M. & Pandey, J. (Eds.). (2008). *Perspectives on learning disabilities in India: Current practices and prospects*. Sage.
Timmons, V., & Alur, M. (2004). Transformational learning: A description of how inclusionary practice was accepted in India. *International Journal of Special Education, 19*(1), 38–48.

UNESCO. (2019). *State of the education report for India 2019: Children with disabilities*. Author.

UNICEF. (2013). *The state of the world's children 2013: Children with disabilities*. Author.

Unni, J. C. (2012). Specific learning disability and the amended "Persons with Disability Act". *Indian Pediatrics, 49*, 445–447. http://medind.nic.in/ibv/t12/i6/ibvt12i6p445.pdf

Vaidya, S. (2020). Disorder, disability, difference: (Re)presenting autism in India. In N. Mehrotra (Ed.), *Disability Studies in India* (pp. 43–59). Springer. https://doi.org/10.1007/978-981-15-2616-9_1

Varma, P. (2007). We need to go back to the drawing board. In A. Kaushik (Ed.), *Shiksha: The challenge of Indian education* (pp. 41–50). Tehelka.

Ware, L. (2010). Disability studies in education. In S. Tozer, B. P. Gallegos, A. Henry, M. B. Greiner, & P. G. Price (Eds.), *Handbook of research in the social foundations of education* (pp. 244–260). Routledge.

Wilkerson, I. (2020). *Caste: The origins of our discontents*. Random House.

World Health Organization/World Bank. (2011). *World Report on Disability 2011*. Author.

Yin, R. K. (2014). *Case study research: Design and methods* (5th ed.). Sage.

CHAPTER 2

The Global Context for Inclusive Education

How did an inclusive program like Education For All create a new group of marginalized students? Part of the answer can be found in our contemporary context of neoliberal globalization and international development. Never before has our world been so enmeshed in terms of commerce and ideas as it is now (Hansen, 2020; Parker, 2010). As Hansen notes, in previous waves of globalization, "ordinary people weren't able to travel virtually anywhere, walk into a store and buy goods from almost any country" (p. 3). With the concomitant need for a common mode of communication and a global workforce, English has become the global *lingua franca* (Bhatia, 2018), and the purpose of education has been re-envisioned. Where education once was an elite bastion, a privilege of birth, it is now the right of all children—who are no longer required/expected to work for a living or contribute to the family income. The ideology of market fundamentalism assumes that students everywhere will learn the same skills needed for this global market (Pritchett, 2013; Stiglitz, 2003), contributing to the notion that learning algebra and English are more valuable for a rural child in Ladakh or a Baiga child in Madhya Pradesh, so they can aspire to a job as, say, a computer programmer or coder, rather than maintain their indigenous knowledge about rice-planting and local medicinal plants (Black, 2011; Sarangapani, 2003). This global milieu and the perception that acquiring

M. Kalyanpur, *Development, Education and Learning Disability in India*, Palgrave Studies in Disability and International Development, https://doi.org/10.1007/978-3-030-83989-5_2

English language skills is an avenue to social and economic mobility have provided the impetus for millions of low-income children in India who speak languages other than English and can receive free government education to pay for private English-medium education (Bhatia, 2018; Chattopadhay & Roy, 2017; EPW Engage, 2021).

The ideology of international development has furthered this agenda by establishing ideals of development "shaped by the geopolitical interests and normative approaches of the predominantly Western hegemons of the world system" (Mundy & Manion, 2015, p. 51). Setting a template for development based on a quality of life that was enabled by colonialism, international aid agencies placed countries on a continuum of development whereby developing countries, the global South, were in deficit (Chang, 2003; Loomba, 2015). The template included universal primary education, what we now call Education For All. The intent is not to decry this outcome—certainly an educated population contributes greatly to the wealth of a nation, but to question the manner and the purpose. As this chapter shows, the assumption of universality, that all countries should follow the same path of development and that this template is applicable across all nations, regardless of their social, political and economic circumstances, embedded in contemporary globalization and international development is problematic. It examines the distinguishing features of contemporary neoliberal globalization and international development that separate it from previous waves of globalization. It identifies the legacies of educational systems from colonial globalization that persist as a result of continuing postcolonial Eurocentric epistemological hegemony in the form of international development. The last section examines the goodness of fit for universal standards of inclusive education for India.

Globalization During Colonialization

Artiles and Dyson define globalization as "the tendency for human activities which occur at places geographically remote from each other nonetheless to interact with each other so that local practices in terms, say, of culture, politics or economics—become overlain, or are entirely swept aside by global patterns" (2005, p. 38). Historians have pointed to earlier eras of globalization, such as the year 1000 (Hansen, 2020) or the early modern age (Parker, 2010), where the emergence of empires led to new forms of interactions. However, our contemporary era of globalization has been the most far-reaching and transformative (Loomba, 2015;

Motha, 2014). For example, commercial networks are more now tightly connected (Hansen, 2020; Parker, 2010). Combined with increased ease and speed of travel for large-scale migration, this has meant that people in one part of the world, say India, are receiving an education that gives them skills such as computer engineering and English language education that they would need to work and live in another part of the world (Stromquist & Monkman, 2015).

Colonial globalization in the early modern age was made possible by government-initiated territorial expansion, strong economic and political protectionism, such as slave-based modes of production and monarchic centralized governance (Parker, 2010). Resulting in the imposition of products and ideas in one, imbalanced direction from Western Europe and North America to the rest of the world, it gave these countries the opportunity to amass great wealth and precipitated the transition to capitalism in Europe (Easterly, 2006; Loomba, 2015). By 1913, at the height of Empire, the West controlled 60% of the world which together generated almost 80% of the wealth.[1] This prosperity contributed gradually to social and political change on what are now universally considered standard indicators of development, such as access to universal health care and education, democratic rule and women's and children's rights (Chang, 2003; Chang & Grable, 2004; Easterly, 2006).

Colonial Globalization and Education

At the beginning of colonialism, the historical tradition that education was the exclusive preserve of only an elite few within societal hierarchies was fairly global. This led to the exclusion of many. For instance, only upper-caste males received an education in India (Ghurye, 1932, as cited in Celarent, 2011); the working class in Europe and racial minorities in the US did not (Pritchett, 2013; Stratton, 2016). The economic and capitalist gains made through colonial plundering were rationalized on an epistemological foundation of the superiority of the Western civilization and the needed justification for civilizing other peoples, what was referred to as "the white man's burden" (Black, 2011). In India, the issue of language convoluted the colonization process further, infiltrating local educational systems (Annamalai, 2004). While the primary purpose of the East India Company was trade, it was administered by representatives of the British crown who saw the advantage in acquiring Persian and Sanskrit toward building relationships with the existing elites and rulers (Chavan, 2013).

However, once power was consolidated, the colonial gaze changed and "colonial education (became) a tool of imperial hegemony" (Upadhyay, 2012, p. 158). Investment was primarily at the level of higher education, as colleges supported by the Crown began to offer curriculum in English to Indians, intended, as stated in the infamous Macaulay Minute of 1835, to "form a class who may be interpreters between us and the millions whom we govern, – a class of persons Indian in blood and colour, but English in tastes, in opinions, in morals and in intellect" (Article 34). Education was structured on the premise that the "native" needed to be civilized, and that English cultural values and culture were superior to that of the Indians, rapidly eroding the indigenous curriculum that had flourished for centuries. The target for this "modern education in English" was "the very small number of upper-class Indians," on the theory of "downward filtration" strongly evocative of today's "trickle-down economics," that the "educated Indians would transmit this knowledge to groups lower to them" (Upadhyay, 2012, p. 162). The educational system of *pathshala* for Hindus and *madrassa* for Muslims, which had been decentralized in curriculum, assessment and selection of teachers under the patronage of the Brahmanic elite and Muslim aristocracy, became highly centralized, dependent on government grants for its survival (Chavan, 2013; Upadhyay, 2012). Since universal education was not a profitable enterprise, elementary school investment was left to missionaries, who set up English-medium schools for children of British administrative officers and Anglo-Indians. In this period, the very architecture of school changed from a *guru* and his pupils gathering in the open air, which was seen as "unwalled and spatially unspecific," to a building consisting of "a closed, four-walled room with wooden desks and benches" (Kumar, 2017, as cited in Upadhyay, 2012).

To suggest that the impositions of colonial globalization were unwelcome, however, is naïve: It overlooks the complexity of the relationship between colonizer and colonized, and conveys an implicit homogeneity of the colonized (Fanon, 2004; Loomba, 2015). As Fanon (2004) pointed out, colonialism was successful not only because of the subjugation of the masses but also because of the co-opting of a comprador class that proceeded to benefit from or take advantage of the situation. The British Empire's decision to introduce English in Indian schools and higher education as a deliberate imperial strategy created a English-speaking, English-educated Indian elite (Chavan, 2013; Kopf, 1984) which continues to maintain its hegemony in contemporary India (Bhatia, 2018; Loomba, 2015).

The next section examines features of postcolonial policies and structures that continue to place the developing world or the global South[2] at a disadvantage and perpetuate impositions from globalization.

Contemporary Globalization and Neoliberal Economics

When colonialism ended in the 1950s and 1960s, the former colonies were as economically devastated as the Western European countries had been at the end of World War II. The idea was conceived that, just like the Marshall plan, which had been so successful in Western Europe's recovery, funds would be diverted to former colonies to help them develop. The United Nations, or the UN, was constructed and Western European countries and the US began to contribute small portions of their GDP to development (Menashy & Manion, 2016; Mundy & Manion, 2015). Since the intention was to help former colonies to develop like the west, UN agencies like UNESCO and UNICEF also developed a network of technical assistance to bridge this "gap" (Menashy & Manion, 2016), consisting of experts from the West who could guide Asian, African and South American countries along this path of development.

This imposition of a new global order or Empire was different from the overt domination of sovereignty of colonialism, in that, as Hardt and Negri (2000) argued, contemporary Empire is imperial and not "imperialist" because it absorbs the global South into a new international network rather than "invade, destroy and subsume subject countries within its sovereignty" (p. 182). However, their argument, with neoliberal scholars (e.g., Friedman, 2005; Tooley, 2009) that this change has the benefit of facilitating resistance while dissolving economic inequities, overlooks the pernicious effects of ideology (Loomba, 2015). Postcolonial scholars argue that hegemonic impositions of ideology and what development means (Loomba, 2015; Mundy & Manion, 2015; Said, 1979; Stromquist & Monkman, 2015; Tikly, 2001; Grosfoguel, 2013) have established a tyranny of the universal (Rao & Kalyanpur, 2020). As a result of the currently prevalent ideology of market fundamentalism where the ideology of capitalism rules supreme (Klees, 2017; Sainath, 2001), "globalization itself has... been disadvantageous to developing countries, especially the poor within those countries" (Stiglitz, 2003). The civilizing mission of colonialism has morphed in contemporary globalization into the holy trinity of development—a free market economy, democratic government and human rights—that has come to be perceived as universal desiderata (Chang, 2003; Leiberman, 2017), despite the inherent contradiction between the embedded values of the common good in democratic governance and of individualism in capitalism.

While these outcomes are laudable, their imposition overlooks significant factors. One, these changes occurred from the excess wealth afforded through colonialism (Loomba, 2015), while these resources are sorely lacking in developing countries and the donor community has failed to provide resources promised to meet these targets (Mundy & Manion, 2015). Two, colonial countries had the benefit of time of some centuries to accomplish these outcomes, enabling them to focus on the most pressing priorities first (Chang, 2003; Chang & Grable, 2004; Easterly, 2006), whereas the Millennium and now the Sustainable Development Goals have targets of 15 year periods for equally prioritized outcomes. Countries are perceived as having failed if these goals are not met (Chabbott, 2015; Chang, 2003). Today's developed countries, as donors, are holding the developing world to standards of political and economic sophistication that they themselves were not subjected to and would not have met at a similar phase in their development (Chabbott, 2015).

More important, these structures emerge from western sensibilities that may not prevail in other contexts. Imposing a "one-size-fits-all" approach (Stiglitz, 2007) fails to recognize local social realities and existing assets.[3] By the end of colonialism, the notions of universal education and the public school became rooted in the global North with a growing sense of nationhood and the need to create a citizenry that shared a similar language and national values (Bellah et al., 2007; Chaterjee, 1993; Stratton, 2016; Tyack, 2003). The role of a child had become that of a student, furthered by the abolition of child labor and the promulgation of the right of a child to education. However, the ostensibly liberatory purpose of education as a great equalizer was undercut by persisting social hierarchies which maintained their cultural hegemony over the curriculum offered in schools (Bourdieu & Passeron, 1990; Collin & Apple, 2010). Assumptions about universality of knowledge continued to privilege elite children who already had educated parents who most likely spoke the same language that was being taught in schools, who had a print-rich home environment and access to books they would most likely encounter in schools—in other words, all the advantages that traditionally disenfranchised children lacked, causing them to be further disadvantaged when presented with an unfamiliar curriculum. These educational inequities continue to beset the global North (Collin & Apple, 2010; Kozleski et al., 2015). Yet this flawed educational system has become part of the rights-based platform of the "international education for development regime" (Mundy & Manion, 2015, p. 51) and the template for universal education globally.

Neoliberal globalization policies instituted since the 1990s have had further deleterious effects on developing economies (Stiglitz, 2007; Stromquist & Monkman, 2015). Countries were urged or moved voluntarily toward free market economies, public–private partnerships were encouraged, and public health and education spending severely reduced and the sectors largely privatized (Chabbott, 2015; Garrett, 2000; Stiglitz, 2007; Stromquist & Monkman, 2015). Garrett (2000) notes of India in 1992, "at the time of record-breaking economic growth, the public health sector was its lowest rank of any major spending category. Just ahead of it was education" (p. 23). This divestment came at the cost of international development policies recommending Education For All (EFA).

Education in Contemporary Neoliberal Globalization

The impetus for universal education became part of the international development agenda. EFA's promise recognized the prevalence of large swathes of school-age populations who were excluded from educational systems, such as girls, working classes and the poor, religious minorities, lower castes and other marginalized groups within the global South (Ajmera, 2016; Miles & Singal, 2010; UNESCO, 2015).

The belief that successful policies and practices in the global North would be equally applicable and successful in the global South pushed "isomorphic mimicry" (Pritchett, 2013) in the uncritical transplanting of ideas or policy borrowing from the global North (Mukhopadhyay & Sriprakash, 2011). This universalization of education resulted in cookie-cutter approaches, or the "McDonaldization of education" Goodley (2011, p. 144), undermining indigenous structures of schooling already considerably weakened by colonialism. Deep-seated notions of school as an architectural location that children go to and sit at desks for a certain number of hours each day each week—the very structure of what being at school means, traveled globally. These notions overlooked the goodness of fit with, say, the lifestyle of children of migrant workers and nomadic cultures, who speak a different language from that taught in the school, or are required to work for a living or in the fields (Ajmera, 2016; Alessa, 2009; Dyer, 2001). They dismantled indigenous practices of learning while under a tree or seated on the floor or imbibed through observing community elders (Alessa, 2009; Kumar, 2007, as cited in Upadhyay, 2012; Sarangapani, 2003; Scribner & Cole, 1973), although "the content, method, and direction of such learning were controlled by tradition and were quite formal" (Miles et al., 2014, p. 343). They overlooked the possibility of a school calendar that responded to agricultural cycles that still prevailed in newly developing countries (Alessa, 2009; Black, 2011; Kalyanpur, 2014) as it had in Europe and the US before shifting to the rhythms of industrialized societies (Stratton, 2016).

Even the purpose of education became universal, with expectations that children study through several years of education before taking a national standardized assessment to receive formal recognition and the economic dividends of their efforts toward creating a citizenry for a workplace with shared values (McDermott et al., 2011; Stromquist & Monkman, 2015; Tikly, 2001). As Menashy and Manion (2016) note, "the importance of education was understood to extend beyond what it meant for the individual to include the social function of education in processes of cultural production and reproduction" (p. 321). In the 1990s, the growing recognition of the importance of English as a global *lingua franca* converged with international aid agencies' recommendations for structural adjustments and neoliberal economic reforms (Mundy & Manion, 2015). Private investment in education and the increasing aspiration to social and economic mobility made low-fee-paying schools attractive to low-income families (Day et al., 2014; Tooley, 2009). In India, these private schools offered English immersion just as the schools for the elite did, although students from low-income backgrounds were unlikely to have access to English in their home environments or immediate communities (Boruah, 2017; Kalia & Reese, 2009). This global template of education was ostensibly intended to create a workforce in any part of the world skilled enough to compete within the world labor market—creating the illusion that a poor, urban child in Mumbai can go to private school and acquire sufficient fluency in English and skills to continue higher education in top tier universities toward ultimately becoming rich (Bhatia, 2018; Black, 2011).

Families who were hitherto denied an education now avidly seek an education for their child, believing in its promise for an aspirational lifestyle and an avenue out of poverty (Singal et al., 2019). This promise is played out on the websites of NGOs highlighting their one or two students who have succeeded in getting a scholarship into a high ranked pre-university college program or through billboards of tutoring programs for school-leaving or college-entrance exam preparation touting their "toppers." In reality, these successes are very few and far between. For many, education has not translated into real change, with acquisition of skills meaningful to their lifestyles and reduction of poverty or social discrimination, and has only succeeded in creating a disconnect with indigenous knowledge and ways of life (Bhatia, 2018; Black, 2011; Mosse, 2018; Pritchett, 2013; Rabukawaqa, 2009). Additionally, recent efforts toward decentralization through neoliberal globalized structural adjustments have reduced government spending on public education and increased private investment in education, resulting in exacerbating further inequities in educational access (Garrett, 2000). When assessments or evaluations do identify these deficiencies in a pyramid system that has room for just a few at the top (Wagner et al., 2018), the blame rarely falls on the *importation* of centralized and institutionalized educational structures that were a poor contextual fit to begin with (Mukhopadhyay & Sriprakash, 2011).

The built-in inequities of current structures have not helped children who have traditionally struggled to access education as this standardization or "global norming" (McDermott et al., 2011) has become even more entrenched since the new millennium. Ineluctably, 121 million school-aged children and adolescents are out of school worldwide, most of them poor, rural and, often girls (Ajmera, 2016; UNESCO, 2015). As Miles, et al. (2014) noted,

> Market-driven, neo-liberal education policies that emphasise excessive accountability, competition and individualism are exclusionary, and threaten educational equity and the inclusion of children with disabilities. (p. 343)

The next section analyzes the challenges of transferability in the evolution of special and inclusive education services in India within contexts of "global inclusionism" (Le Fanu, 2013) and EFA.

Challenges in Transferability of Inclusive Education as a Universal Standard

Education for children with disabilities in the global North began as segregated, special schools, a parallel structure that helped develop a robust corpus of special education and rehabilitation services professionals and curricular resources, occurring as recently as the 1970s in the US with the enactment of the historic Education of the Handicapped Act (EHA) in 1975 (Leach & Helf, 2016). The movement toward inclusion began in the 1990s with growing perceptions that special schools maintained discriminatory segregation. The Regular Education Initiative, which re-conceptualized the role of general education teachers as being pivotal to responding to the needs of students with disabilities in general education classrooms, with support from special education personnel (Leach & Helf, 2016) was incorporated into EHA in 1990, which was renamed the Individuals with Disabilities Education Act (IDEA). This became the universal template for inclusive education and the impetus for global norming (McDermott et al., 2011). Encapsulated in "virtually the Bible of the inclusion movement" (Urwick & Elliott, 2010, p. 139), the 1994 Salamanca Statement and Framework for Action on Special Needs Education[4] declared that:

> (Children) with special educational needs must have access to regular schools which should accommodate them within a child-centred pedagogy capable of meeting these needs, (as) regular schools with this inclusive orientation are the most effective means of combating discriminatory attitudes, creating welcoming communities, building an inclusive society... and improv(ing) the efficiency and ultimately the cost-effectiveness of the entire education system. (pp. 7–8)

The evolution of special and inclusive education services in contemporary India has followed a path very similar to that in the global North, both programmatically and legislatively (Singal & Jeffery, 2015). As a recent report from the National Law University states:

> The concept of inclusive education has not originated in India with NEP (National Education Policy), rather it is inspired by western cultures like that of the UK and USA where there are inclusive schools. The very idea of inclusive schooling is very pious and the intent of legislature stands out to be bona fide. But, how far does this importation of western ideas without

required changes to suit the dynamics of Indian society tenable in terms of working is a question that needs to be pondered upon. (Pro Bono and Legal Aid Clinic on Education & Legal Services Committee, 2021, p. 81)

Indeed, the issues of "where," or the appropriateness of types of placements for children with disabilities, and the "who," or which children are deemed disabled, have raised concerns about the goodness of fit. For instance, given that so many other student populations were marginalized in India, when initially implemented, inclusion referred to an education not merely for children with disabilities but for all disadvantaged groups (Singal, 2005, 2006).

The Problem of Where: Special Schools or Inclusion?

Services began with the establishment of special schools in alignment with "reports of its success in the West" (Tillou, 1972, as cited in Kalyanpur, 1996, p. 253), initiated by Christian missionaries, voluntary agencies and later parent organizations. Although these were few in number and served a sliver of the population of children with disabilities, mostly urban and middle class, the trend toward special schools was clear. Over the next two decades, the number of special schools grew as the government both encouraged and supported voluntary effort. The government also investing its own funds to establish national institutes for specific categories of disabilities in urban centers to train teachers and conduct research, among other activities.

In the 1970s, a series of studies conducted by UNESCO (1973, 1977, 1978, as cited in Kalyanpur, 1996) on educational services for children with disabilities in different countries, concluded that, for social and economic reasons, integrated education was more effective than special education. They recommended that national policies in developing countries be directed to providing equal access to education for all. In response, the Indian government initiated the Integrated Education for Disabled Children (IEDC) as a pilot program in Delhi toward including children with physical disabilities,[5] and later extended to include children with severe disabilities and children with learning disabilities, although no details were provided on what was understood by the term "learning disability" or what supports were available (National Council for Educational Research and Training, 1987).[6] The first concerns about the importation of Western curricula and instructional materials, teaching techniques and organizational structures also began to emerge (Baine, 1988; Miles, 1981, 1985). Under criticism that the largest package of health and education services for children and women in rural areas, the Integrated Child Development Scheme (ICDS) established in 1974 did not include children with disabilities among its beneficiaries (Alur, 2002), eventually the IEDC program was subsumed under generic rural development and poverty reduction programs.

Propelled by increased funding from the World Bank, the Indian government also expanded access to schooling under the Janshala Schools EFA program (1998–2004). It created awareness about the importance of education in low enrollment districts and offered free education to "integrable" children with disabilities in regular classrooms at government-funded schools with transportation and assistive devices (Gandhe, 2004). In 2003, the newest and most expansive EFA initiative, Sarva Shiksha Abhiyan (SSA),[7] promised to accommodate other students with disabilities within alternative and non-formal schooling systems and provide residential and non-residential "bridge courses" or transitional programs for mainstreaming to children with disabilities who had never been to school.

The government also promulgated several disability-related legislations.[8] In 2016, it enacted the Rights of Persons with Disabilities (RPD) Act, amended from the Persons with Disabilities (Equal Opportunities, Protection of Rights and Full Participation) Act of 1995. This comprehensive legislation gives persons with disabilities until 18 years of age the right to education (Ministry of Law & Justice, 2016) and strongly contains elements of the 2008 UN Convention on the Rights for People with Disabilities (CRPD) and the 1990 US Americans with Disabilities Act. For instance, it introduces the concept of "reasonable accommodation." However, the term "accommodation" in India is associated most closely with its meaning of a lodging or a residence (as in "Is the accommodation you are living in now reasonable?"), and this new technical meaning of modifications, such as ramps or menus in Braille, that an agency or company can provide within reasonable financial costs to increase access to people with disabilities was not immediately apparent to all stakeholders. Continuing with the westward-looking stance on pedagogy, the RPD Act mandates use of Universal Design and assistive devices.[9] Most significantly, the legislation uses terminology in line with the UNCRPD and the Salamanca Statement to clarify that inclusive education means "a system of education wherein students with and without disability learn together and the system of teaching and learning is suitably adapted to meet the learning needs of different types of students with disabilities." Yet, it allows possibilities for a range of educational settings, stating that "every child with benchmark disability… shall have the right to free education in a neighbourhood school, or in a special school, of his choice."

As Mukhopadhyay and Sriprakash (2011) noted, policy borrowing "located in global agendas of education development" (p. 322) becomes policy translation where they are "re-worked, re-interpreted, and re-enacted contextually" (p. 323). They add:

> Such translations of policies, within a confluence of differently situated actors and their interests, are likely to be particularly marked in countries such as India where the polity is significantly heterogeneous in terms of social contexts and institutional positions. (p. 323)

Scholars have argued that transposing the strong rights-based approach of "the international vision for inclusion" within the context of "the pressures of neoliberalism and globalisation, alongside meeting international development goals" creates "tensions, contradictions and dilemmas" (Miles et al., 2014, pp. 339–340). Some suggest that inclusive education should be more broadly defined to include a variety of settings (Alborz et al., 2013; Urwick & Elliott, 2010), arguing that "inclusive education as a practice is definitely contextual," advocating that "it should be a matter of choice" with the need to consider the significance of special schools (Chennat, 2019, p. 45). Indeed, within the Indian context, inclusion has been "translated" (Mukhopadhyay & Sriprakash, 2011) as any educational service, even a special school, for children with disabilities because it represents an immense step toward their inclusion in society (Kalyanpur, 2008b). Further, the argument for inclusion as a right of children with disabilities to an education is perceived to be disconnected from the realities of local contexts of limited resources: In Mumbai alone, there are only 47 special education professionals for 4,700 municipal schools (Limaye, 2019). As Chennat (2019c) concluded:

> Special schools do have a significant position in the continuum of educational settings for children with disabilities. Children and parents should be able to make informed choices of the right educational setting for their child from a variety of settings. (pp. 45–46)

The parallel co-existence of special schools is often perceived as necessary, given that children with disabilities have been the last to be included in schools and least likely to be included at all (World Health Organization/World Bank, 2011). In India, the National Census on people with disabilities conducted in 2002 found that, despite all efforts, about 94% of children with disabilities did not receive any educational services, and some groups, such as children in rural areas, girls with disabilities and individuals with mental retardation, were further under-represented (National Sample Survey Organisation, 2003, as cited in Kalyanpur, 2008a). More recently, UNESCO (2019) stated that over 28% of children with disabilities in India continue to be out of school, much higher than the national estimate (2.97%) of total out-of-school children.

The Problem of Who: Identification and Labeling

In line with World Health Organization's International Classification of Functioning, the RPD legislation distinguishes between disabilities and benchmark disability[10] and states the need to "designate persons, having requisite qualifications and experience, as certifying authorities, who shall be competent to issue the certificate of disability." This distinction between "real disability" and "functional disability" has been controversial for its dependence on the decision-making authority of professionals who would be making the determination on a person's capacity (Florian et al., 2006) and for maintaining a strong medical focus. For instance, the Chairperson of the Medical Council of India is a member of the national-level Central Advisory Board on Disability. Only government-run clinics have the authority to issue disability certificates for learning disabilities, while private clinics issue opinion certificates that include recommendations for teaching strategies and benchmark disabilities must be re-certified on an annual basis (Ahmad, 2015; Unni, 2012). The United Nations Economic and Social Commission for Asia and the Pacific (UNESCAP, 2008) itself has recommended conducting ethnographic studies as a precursor to household surveys for disability statistics to ensure that the determination of the presence of a disability is based more appropriately on indigenous knowledge rather than foreign or external perceptions. However, an adherence to universal standards does not allow for indigenous interpretations or local conditions to impact on culturally responsive treatment alternatives (e.g., Florian et al., 2006; Harry, 2020; Mukherjee, 2016).

The similarity in definitions of specific learning disability in the US and in India is strongly indicative of a borrowed concept. As defined in the RPD Act (Part II, D 2a), specific learning disabilities "means a heterogeneous group of conditions wherein there is a deficit in processing language, spoken or written, that may manifest itself as a difficulty to comprehend, speak, read, write, spell, or to do mathematical calculations and includes such conditions as perceptual disabilities, dyslexia, dysgraphia, dyscalculia, dyspraxia and developmental aphasia." This is very similar to the US definition.[11] However, there are two significant divergences. One, the US definition includes a caveat that factors such as linguistic diversity as possibly confounding a student's academic struggle be eliminated before making a determination about the presence of LD, which is not included in RPD's definition.[12] Scholars have expressed concern over this omission, given India's multilingual context (e.g., Karanth, 2008). For instance, recent research shows that low-income students at English-medium schools are almost twice as likely to be identified as "at risk" of learning disabilities than elite students, suggesting that language may be a factor in their academic struggles (Shenoy & Wagner, 2019). Two, although scholars have noted the inclusion of specific learning disabilities as progressive (Ahmad, 2015; Unni, 2012), it is listed under intellectual disabilities with autism spectrum disorders, unlike in the US, where it is a standalone category. This second concern may relate to perceptions and prevalence rates. In the US, students with LD constitute over 50% of all students with disabilities and is a high-incidence category; in India, students with LD constitute 7% of all children with disabilities (Government of India Census, 2011, as cited in UNESCO, 2019) and are still considered "one of the high incidence disabilities" (Panshikar, 2019, p. 175). As Chapters 5 and 6 describe, these reworked translations lend themselves to elements of confusion about who students with learning disability in India truly are.

Conclusion

The colonial mindset of deficit that justified its "civilizing mission" continues to prevail in our contemporary era of neoliberal economics and market fundamentalism. Postcolonial globalization maintains this belief through the concept of development, and the expectation that the global South achieve similar goals of progress as the global North, and that if it does not, it has somehow failed. However, this expectation does not take into account the severe depletion of resources in the global South from colonialism and that these indicators of development took the global North several centuries to attain, but are being set to be met in a few decades for the South. It also overlooks the question of compatibility with models already problematic in the global North being transposed to contexts with very different local realities.

Pressure from the international development sector to implement universal access to education, as one of the indicators of development and neoliberal economic reforms to create a global workforce that speaks English, the global *lingua franca*, has provided the impetus for large-scale expansion of educational services for all. Universal standards, like the 1994 Salamanca Statement, have urged inclusive educational models, and private investment in schools has established immersion English models, regardless of their alignment with and relevance to local realities. In responses to these challenges of importation as an ongoing motif in the provision of services for children with disabilities in second-generation countries, many urge that stakeholders, including international organizations, consider local contexts and "indigenize" the framework to make it more suitable for these local conditions (e.g., Jelas & Ali, 2014; Wang et al., 2017). As Maudslay (2014) puts it:

> If, even in countries where notions of inclusive education first emerged, the pathway to implementation is far from straightforward, it is likely that the complexities and dilemmas which need to be explored and worked through are far larger in many Southern countries and also that the solutions might be significantly different.... a top-down donor-led model of inclusion can end up not leaving any space for the study of the reality upon which educational changes must be based. (p. 423)

The next chapter presents a deeper analysis of these local realities and the contextual factors at the meso or national level that contribute to this misalignment.

Notes

1. In India, for instance, the British colonialists set up an extensive railroad network for the express and sole purpose of transporting raw goods from the hinterland to the port cities. Raw cotton, which had previously been converted into cloth in Indian mills,

was now being shipped to mills in England and the manufactured cloth shipped back to be sold at higher tariffed rates in the Indian market. Trading arrangements like this had a deleterious effect on India's economy: Around 1750, prior to colonization, India contributed 24% of the world's Gross Domestic Product (GDP) while Great Britain's contribution was 1.8%. As a result of the transfer of resources from India to Britain, on the eve of India's independence from Britain in 1947, India's contribution to world GDP was 1.8% while Great Britain's was about 25% (Maddison, 2004).

2. There is need for a caveat on the binary implied in the interchangeable terms west/developed countries/global North, which tend to be used in contrast to the terms east/developing countries/global South. The former set of terms originally referred primarily to the group of countries, such as Europe and the US, which having benefitted from colonialism as colonizers, are donors with the international aid network, and now benefit the most from contemporary globalization, their geographical positioning within the western hemisphere contributing to the notion of the "west." However, over time, as other countries, such as Australia, New Zealand, Singapore and Japan, have also met the economic and social indicators of development and benefitted from contemporary globalization, this demarcation has been expanded beyond mere geography. Additionally, not all former colonies are recipients of aid, such as Brunei, while some provide aid, such as South Korea and Taiwan; on the other hand, some countries that have not been colonies receive aid, such as Nepal. Finally, as Kozleski et al. (2015) point out, this disadvantage is also experienced intra-nationally; for instance, in the resource-rich US, which constitutes part of the global North, students from culturally and linguistically diverse and low socio-economic backgrounds face similar barriers to accessing inclusive education to students in the global South, while students from upper caste and high socio-economic backgrounds in India, a poorly resourced country in the global South, enjoy privileges similar to advantaged populations in the global North. They suggest an alternative binary of first-generation inclusive education for contexts where the concept and practice first emerged and second-generation inclusive education for contexts where the concept and practices have been transferred to or implemented in

a second iteration. Thus, the terms "global North/South" and "first/second generation inclusive education" countries refer not to a geographical delineation of space, but a historical and sociological context that has traditionally excluded and disadvantaged some countries and certain communities of people.

3. For instance, the western notions of democracy with its underlying values of meritocracy, individualism and human rights are considered superior to the patron politics and nepotism that prevail in many developing countries, even though these structures are often based on longstanding traditions (Béteille, 2012; Samarendra, 2011). In India, patronage and caste politics has contributed as much to the advancement of *Dalits* and lower castes as the government policies of reservation quotas, since the entrenched nature of upper-caste/class interests perpetuates a "sticky floor effect" that limits economic and social mobility for the poorest of the poor (Béteille, 2012; Bhatia, 2018; Loomba, 2016; Mosse, 2018).
4. Since then, the International Classification of Functioning developed by the World Health Organization (WHO, 2011) and the 2007 UN Convention on the Rights of Persons with Disabilities (UNCRPD) have been added to the international standards.
5. Specifically "blind and partially sighted children, deaf and partially hearing children, orthopaedically and neurologically handicapped children, and children with multiple handicaps" (Ministry of Social Welfare, no date, cited in Kalyanpur, 1996, p. 257).
6. The program provided funding for training teachers, and for equipment, books and assessment facilities, and recommended that materials, like Braillers and personnel, such as itinerant teachers and consultants be consolidated in Project Area Resource Centers. Although the IEDC was expanded as a nationwide project in 1987, at its height in 1994 with full UNICEF funding, it was implemented in just ten out of 29 states (Rao et al., 2005). An evaluation of the original pilot IEDC program found that, as a result of inadequate dissemination and resource allocation, mostly children with visible disabilities, such as motor or visual impairments, were identified and integrated, few parents of children with disabilities knew about the project, and few children actually received any equipment or services, while the lack of training bred teachers who were unwilling or unable to respond effectively to their students with disabilities (Seetharam, 1982, cited in Kalyanpur, 1996).

7. Now called Samagra Shiksha Abhiyan.
8. In 1992, responding to the need for capacity building, the Rehabilitation Council of India (RCI) Act mandated minimum standards for training and teacher certification for professionals in the field of special education and rehabilitation (Misra, 2000). In 1999, responding to pressure from parent organizations, the National Trust (for the Welfare of Persons with Autism, Cerebral Palsy, Mental Retardation and Multiple Disabilities) Act of 1999 was passed.
9. Specific measures to promote and facilitate inclusive education include, among others, (a) conducting a survey of school going children in every five years for identifying children with disabilities, ascertaining their special needs and the extent to which these are being met, and (b) training professionals and staff to support inclusive education at all levels of school education. In conjunction with the National Education Policy (NEP, 2020), it recommended the establishment of a national repository of high-quality resources on foundational literacy and numeracy will be made available on the National Teacher's Portal, i.e., the Digital Infrastructure for Knowledge Sharing or DIKSHA.
10. It notes that a "person with benchmark disability" means a person with not less than forty percent of a specified disability where specified disability has not been defined in measurable terms and includes a person with disability where specified disability has been defined in measurable terms, as certified by the certifying authority, whereas a "person with disability" means a person with long term physical, mental, intellectual or sensory impairment which, in interaction with barriers, hinders his full and effective participation in society equally with others.
11. Specific learning disability in the US is defined in the Individuals with Disabilities Education Act (IDEA) as "a disorder in one or more of the basic psychological processes involved in understanding or in using language, spoken or written, that may manifest itself in the imperfect ability to listen, think, speak, read, write, spell, or to do mathematical calculations, including conditions such as perceptual disabilities, brain injury, minimal brain dysfunction, dyslexia, and developmental aphasia" (IDEA Sec. 300.8 (c) (10), 2014).
12. Further, in the RPD Act, "specific learning disabilities" is listed under intellectual disabilities with autism spectrum disorders.

References

Ahmad, F. K. (2015). Exploring the invisible: Issues in identification and assessment of students with learning disabilities in India. *Transcience: A Journal of Global Studies, 6*(1), 91–107.

Ajmera, M. (2016). *Invisible children: Reimagining international development at the grassroots*. Palgrave Macmillan.

Alborz, S. R., & Miles, S. (2013). Establishing the foundations for an inclusive education system in Iraq reflection on findings from a nationwide survey. *International Journal of Inclusive Education, 17*(9), 965–987. https://doi.org/10.1080/13603116.2012.725776

Alur, M. (2002). "They did not figure": Policy exclusion of disabled people in India. *International Journal of Inclusive Education, 6*, 101–112.

Alessa, L. N. (2009). What is truth? Where science and traditional knowledge converge. In M. S. T. Williams (Ed.), *The Alaska native reader: History, culture, politics* (pp. 246–251). Duke University Press.

Annamalai, E. (2004). Nativization of English in India and its effect on multilingualism. *Journal of Language and Politics, 3*(1), 151–162.

Artiles, A., & Dyson, A. (2005). Inclusive education in the globalization age: The promise of comparative culturalhistorical analysis. In A. Artiles & A. Dyson, (Eds.), *Contextualizing inclusive education: Evaluation old and new international perspectives* (pp. 37–62). Routledge.

Baine, D. (1988). *Handicapped children in developing countries: Assessment, curriculum, and instruction*. Vector Press.

Bellah, R. N., Madsen, R., Sullivan, W. M., Swidler, A., & Tipton, S. M. (2007). *Habits of the heart: Individualism and commitment in American life*. University of California Press.

Béteille, A. (2012). *Caste and the citizen*. Oxford University Press.

Bhatia, S. (2018). *Decolonizing psychology: Globalization, social justice, and Indian youth identities*. Oxford University Press.

Black, C. (Director). (2011). *Schooling the world: The white man's last burden* [Film]. Lost People Films. https://schoolingtheworld.org/film/

Boruah, P. (2017). Learning English in a low-cost semi-urban English-medium school in India: Challenges, interaction patterns and domains of use. In H. Coleman (Ed.), *Multilingualisms and development* (pp. 289–306). British Council.

Bourdieu, P. & Passeron, J. C. (1990). *Reproduction in education, society and culture*. Sage.

Celarent, B. (2011). Caste and Race in India by G. S. Ghurye. *American Journal of Sociology, 116*(5), 1713–1719.

Chabbott, C. (2015). *Institutionalizing health and education for all: Global goals, innovations and scaling up*. Teachers College Press.

Chang, H. J. (2003). *Kicking away the ladder: Development strategy in historical perspective*. Anthem Press.

Chang, H., & Grabel, I. (2004). *Reclaiming development: An alternative economic policy manual*. Palgrave Macmillan.

Chaterjee, P. (1993). *The nation and its fragments: Colonial and postcolonial histories*. Princeton University Press.

Chattopadhay, T., & Roy, M. (2017). *Low-Fee Private Schools in India: The Emerging Fault Lines* (Working Paper 233). National Center for the Study of Privatization in Education. Teachers College, Columbia University. http://ncspe.tc.columbia.edu/center-news/low-fee-private-schools-in-india/

Chavan, D. (2013). *Language politics under colonialism: Caste, class and language pedagogy in Western India*. Cambridge Scholars Publishing.

Chennat, S. (2019). Conceptualizing disability inclusion. In S. Chennat (Ed.), *Disability inclusion and inclusive education*. (pp. 39–62) Springer. https://doi.org/10.1007/978-981-15-0524-9_3

Collin, R., & Apple, M. (2010). Globalization, schooling, and literacies in an unsettled age. In S. Tozer, B. P. Gallegos, A. Henry, M. Bushnell Greiner, & P. Groves Price (Eds.), *Handbook of research in the social foundations of education*(pp. 296–307). Routledge.

Day, L. C., McLoughlin, C., Aslam, M., Engel, J., Wales, J., Rawal, S., Batley, G., Kingdom, S., Nicolai, S., & Rose, P. (2014). *The role and impact of private school in developing countries: A rigorous review of the evidence*. Department for International Development.

Dyer, C. (2001). Nomads and education for all: Education for development or domestication? *Comparative Education, 37*(3), 317–327. https://doi.org/10.1080/03050060120067802

Easterly, W. (2006). *The white man's burden: Why the west's efforts to aid the rest have done so much ill and so little good*. Penguin.

EPW Engage. (2021, April 23). *English language education in India: How aspirations for social mobility shape pedagogy*. https://www.epw.in/engage/article/english-language-education-india-aspirations-pedagogy

lorian, L., Hollenweger, J., Simeonsson, R. J., Wedell, K., Riddell, S., Terzi, L., & Holland, A. (2006). Cross-cultural perspectives on the classification of children with disabilities: Part I. Issues in the classification of children with disabilities. *The Journal of Special Education, 40*(1), 36–45.

Fanon, F. (2004). *The wretched of the earth*. Grove Press.

Friedman, T. L. (2005). *The world is flat: A brief history of the twenty-first century*. Farrar, Straus & Giroux.

Garrett, L. (2000). Filth and decay: Pneumonic plague hits India and the world ill responds. In L. Garrett (Ed.), *Betrayal of trust: The collapse of global public health* (pp. 15–48). Hyperion.

Gandhe, S. K. (2004). *External evaluation of Janshala programme: synthesised report* (Pune, Indian Institute of Education).

Goodley, D. (2011). Education: Inclusive disability studies. In D. Goodley (Ed.), *Disability studies: An interdisciplinary introduction* (pp. 139–156). Sage.

Ghurye G. S. (1932). *Caste and race in India*. Kegan Paul, Trench, Trübner.

Grosfoguel, R. (2013). The structure of knowledge in westernized universities epistemic racism/sexism and the four genocides/epistemicides of the long 16th century. *Human Architecture: Journal of the Sociology of Self-Knowledge, 11*(1), 73–90.

Hansen, V. (2020). *The year 1000: When explorers connected the world – and globalization began*. Scribner.

Hardt, M., & Negri, A. (2000). *Empire*. Harvard University Press.

Harry, B. (2020). *Childhood disability, advocacy, and inclusion in the Caribbean: A Trinidad and Tobago case study*. Springer.

Individuals with Disabilities Education Act (IDEA). (2014) (Sec. 300.8 (c) (10). https://sites.ed.gov/idea/regs/b/a/300.8/c/10.

Jelas, Z. M., & Ali, M. M. (2014). Inclusive education in Malaysia: Policy and practice. *International Journal of Inclusive Education, 18*(10), 991–1003. https://doi.org/10.1080/13603116.2012.693398

Kalia, V., & Reese, E. (2009). Relations between Indian children's home literacy environment and their English oral language and literacy skills. *Scientific Studies of Reading, 13*(2), 122–145. https://doi.org/10.1080/10888430902769517

Kalyanpur, M. (1996). The influence of western special education on community-based services in India. *Disability & Society, 11*(2), 249–270.

Kalyanpur, M. (2008a). The paradox of majority under-representation in special education in India: Constructions of difference in a developing country. *Journal of Special Education, 42*(1), 55–64.

Kalyanpur, M. (2008b). Equality, quality and quantity: Challenges in inclusive education policy and service provision in India. *International Journal of Inclusive Education, 12*(3), 243–263.

Kalyanpur, M. (2014). Distortions and dichotomies in inclusive education for children with disabilities in Cambodia in the context of globalization and international development. *International Journal of Disability, Development and Education, 61*(1), 80–94. 15p. DOI: https://doi.org/10.1080/1031912X.2014.878546.

Karanth, P. (2008). Learning disability and language learning. In K. Thapa, G. M. van der Aalsvoort, & J. Pandey (Eds.), *Perspectives on learning disabilities in India: Current practices and prospects* (pp. 80–96). Sage.

Klees, S. J. (2017, August 3). *Financing Universal Education*. https://www.project-syndicate.org/commentary/financing-universal-education-by-steven-j--klees-2017-08

Kopf, D. (1984). Orientalism and the Indian educated elite. In G. P. Kelly & P. G. Altbach (Eds.), *Education and the Colonial Experience* (pp. 117–136). Transaction Books.

Kozleski, E. B., Artiles, A. J., & Waitoller, F. R. (2015). Introduction: Equity in inclusive education. In A. J. Artiles, E. B. Kozleski, & F. R. Waitoller (Eds.), *Inclusive education: Examining equity on five continents* (pp. 1–14). Harvard Education Press.

Kumar, N. (2017). The technology and politics of teaching English. In M. J. Tiwary, S. Kumar, & A. K. Mitra (Eds.), *Dynamics of inclusive classrooms: Social diversity, inequality and school education in India* (pp. 183–202). Orient BlackSwan.

Le Fanu, G. (2013). Reconceptualising inclusive education in international development. In L. Tikly & A. Barrett A (Eds.), *Education, quality and social justice in the global south* (pp. 40–55). Routledge.

Leach, D., & Helf, S. (2016). Revisiting the regular education initiative: Multi-tiered systems of support can strengthen the connection between general and special education. *Journal of the American Academy of Special Education Professionals* (pp. 116–124).

Leiberman, R. H. (Director). (2017). *Angkor awakens: A portrait of Cambodia.* [Film.] PhotoSynthesis Productions.

Limaye, S. (2019). Unheard voices: Schooling experiences of parents having children with disabilities in India. In M. J. Schuelka, C. J. Johnstone, G. Thomas, & A. J. Artiles (Eds.), *The SAGE Handbook of diversity and inclusion in education* (pp. 557–570). Sage.

Loomba, A. (2015). *Colonialism/Postcolonialism* (3rd ed.). Routledge.

Loomba, A. (2016). The everyday violence of caste. *College Literature, 43*(1), 220–225.

Maddison, A. (2004). *The world economy, volume 2: Historical statistics.* OECD Development Centre.

Maudslay, L. (2014). Inclusive education in Nepal: Assumptions and reality. *Childhood, 21*(3), 418–424. https://doi.org/10.1177/0907568213514778

McDermott, R., Edgar, B., & Scarloss, B. (2011). Global norming. In A. Artiles, E. Kozleski, & F. Waitoller (Eds.), *Inclusive education in five continents* (pp. 223–235). Harvard Education Press.

Menashy, F., & Manion, C. (2016). The historical evolution and current challenges of the United Nations and global education policy-making. In K. Mundy, A. Green, B. Lingard, & T. Verger (Eds.), *Global policy and policy-making in education* (pp. 319–334). Wiley Blackwell.

Miles, M. (1981). *A short circuit of mental handicap in India with some sparks of development.* [Monograph]. Mission Hospital.

Miles, M. (1985). *Where there is no rehab plan: A critique of WHO's scheme for community based rehabilitation.* [Monograph]. Hesperian Foundation.

Miles, S., & Singal, N. (2010). The Education for all and inclusive education debate: Conflict, contradiction or opportunity? *International Journal of Inclusive Education, 14*(1), 1–15.

Miles, S., Lene, D., & Merumeru, L. (2014). Making sense of inclusive education in the Pacific region: Networking as a way forward. *Childhood, 21*(3), 339–353. https://doi.org/10.1177/0907568214524458

Ministry of Law and Justice. (2016). *The rights of persons with disabilities Act*. Legislative Department. http://legislative.gov.in/actsofparliamentfromtheyear/rights-persons-disabilities-act-2016

Misra, A. (2000). Special education in India: Current status and future directions. *Division of International Special Education Services Journal, 3*, 6–11.

Mosse, D. (2018). Caste and development: Contemporary perspectives on a structure of discrimination and advantage. *World Development, 110*, 422–436. https://doi.org/10.1016/j.worlddev.2018.06.003

Motha, S. (2014). *Race, empire, and English language teaching: Creating responsible and ethical anti-racist practice*. Teachers College Press.

Mukherjee, S. (2016). *The gene: An intimate history*. Scribner.

Mukhopadhyay, R. ,& Sriprakash, A. (2011). Global frameworks, local contingencies: Policy translations and education development in India. *Compare: A Journal of Comparative and International Education, 41*(3), 311–326.

Mundy, K., & Manion, C. (2015). The education for all initiative. In E. Unterhalter & T. McCowan (Eds.), *Education and international development: Practice, policy and research* (pp. 49–68). Bloomsbury.

National Council of Educational Research and Training [NCERT]. (1987). *Project Integrated Education for the Disabled: a document*. Author.

Panshikar, A. (2019). Specific learning disability: A hidden disability. In S. Chennat (Ed.), *Disability inclusion and inclusive education*. (pp. 175–193). Springer. https://doi.org/10.1007/978-981-15-0524-9_9

Parker, C. (2010). *Global interactions in the early modern age, 1400–1800*. University Press.

Pro Bono and Legal Aid Clinic on Education & Legal Services Committee. (2021). *Inclusivity vis-à-vis National Education Policy 2020*. National Law University. https://nludelhi.ac.in/download/publication/Inclusivity%20vis%20a%20vis%20NEP%202020.pdf

Pritchett, L. (2013). *The rebirth of education: Schooling ain't learning*. Brookings Institution Press.

Rabukawaqa, E. (2009). Opening address. In P. Puamau & F. Pene (Eds.), *Inclusive Education in the Pacific* (The PRIDE Project; Pacific Education Series No. 6, pp. 1–8). Institute of Education, University of the South Pacific. http://www.usp.ac.fj/fileadmin/files/Institutes/pride/Workplans_and_Reports/Paper_and_publications/IEbook.pdf

Rao, S., & Kalyanpur, M. (2020). Universal notions of development and disability: Towards whose imagined vision? *Disability and the Global South, 7*(1), 1830–1851.

Rao, L. G., Narayan, J., & Mani, M. N. G. (2005). *Status of education of children with disabilities.*

Said, H. (2018). The role of English education in post-colonial Egypt: Criticisms and solutions for the future. In G. J. Dei & C. Jaimungal (Eds.), *Indigeneity and decolonial resistance: Alternatives to colonial thinking and practice* (pp. 125–139). Myers Education Press.

Sainath, P. (2001). *And then there was the market.* (Blog). https://www.india-seminar.com/2001/497/497%20p.%20sainath.htm

Samarendra, P. (2011). Census in Colonial India and the Birth of Caste. *Economic and Political Weekly, 46*(33), 51–58.

Sarangapani, P. M. (2003). Indigenising curriculum: Questions posed by Baiga *vidya. Comparative Education, 39*(2), 199–209.

Scribner, S., & Cole, M. (1973). Cognitive consequences of formal and informal education. *Science, 182*(4112), 553–559.

Shenoy, S., & Wagner, R. K. (2019). Language and literacy practices that influence bilingual and bi-literate acquisition in L1 Kannada and L2 English in Bangalore, India. In R. M. Joshi & C. McBride (Eds.), *Handbook of literacy in Akshara orthography* (pp. 373–388). Springer. https://doi.org/10.1007/978-3-030-05977-4_19

Singal, N. (2005). Mapping the field of inclusive education: A review of the Indian literature. *International Journal of Inclusive Education, 9*, 331–350.

Singal, N. (2006). Inclusive education in India: International concept, national interpretation. *International Journal of Disability, Development and Education, 53*, 351–369.

Singal, N., & Jeffery, R. (2015). Inclusive education in India: The struggle for quality in consonance with equity. In A. J. Artiles, E. B. Kozleski, & F. R. Waitoller (Eds.), *Inclusive education: Examining equity on five continents* (pp. 161–183). Harvard Education Press.

Singal, N., Johansson, S. T., & Lynch, P. (2019). Education of children with disabilities: Changing landscape of new opportunities and challenges. In N. Singal, S. T. Johansson, & P. Lynch (Eds.), *Education and disability in the global South: New perspectives from Africa and Asia* (pp. 1–18). Bloomsbury Academic.

Stiglitz, J. E. (2003). *Globalization and its Discontents.* Norton.

Stiglitz, J. E. (2007). *Making globalization work: The next steps to social justice.* W.W. Norton.

Stratton, C. (2016). *Education for empire: American schools, race, and the paths of good citizenship.* University of California Press.

Stromquist, N. P., & Monkman, K. (2015). Defining globalization and assessing its implications for knowledge and education, revisited. In N. P. Stromquist & K. Monkman (Eds.), *Globalization and education: Integration and contestation across cultures* (2nd ed., pp. 1–20). Rowman & Littlefield.

The NEP is the best thing that the Modi government has done since *Swachh Bharat*. (2020, August 23). *The Indian Express*. https://indianexpress.com/article/opinion/columns/education-minister-ramesh-pokhriyal-nep-tavleen-singh-6565927/

Tikly, L. (2001). Globalisation and education in the postcolonial world: Towards a conceptual framework. *Comparative Education, 37*(2), 151–171.

Tooley, J. (2009). *The beautiful tree: A personal journey into how the world's poorest people are educating themselves*. Penguin.

Tyack, D. (2003). *Seeking common ground: Public schools in a diverse world*. Harvard University Press.

Unni, J. C. (2012). Specific learning disability and the amended "Persons with Disability Act". *Indian Pediatrics, 49*, 445–447. http://medind.nic.in/ibv/t12/i6/ibvt12i6p445.pdf

United Nations Economic and Social Commission for Asia and the Pacific (UNESCAP). (2008). *Training manual for disability statistics*. Author.

UNESCO. (2015). *Fixing the broken promise of Education for All: Findings from the global initiative on out-of-school children*. UNESCO Institute for Statistics.

UNESCO. (2019). *State of the education report for India 2019: Children with disabilities*. New Delhi: Author.

Upadhyay, S. B. (2012). Dalits and the modern education in colonial India. In C. Sleeter, S. B. Upadhyay, A. K. Mishra, & S. Kumar (Eds.), *School education, pluralism and marginality: Comparative perspectives* (pp. 152–180). Orient Blackswan.

Urwick, J., & Elliott, J. (2010). International orthodoxy versus national realities: Inclusive schooling and the education of children with disabilities in Lesotho. *Comparative Education, 46*(2), 137–150. https://doi.org/10.1080/03050061003775421

Vagh, S. B., & Sharma, D. (2018). Localized assessments can enhance our understanding of learning at the bottom of the pyramid: Insights from India. In D. A. Wagner, S. Wolf, & R. F. Boruch (Eds.), *Learning at the bottom of the pyramid: Science, measurement, and policy in low-income countries* (pp. 121 137). International Institute for Educational Planning.

Wang, Y., Mu, G. M., & Zhang, L. (2018). Chinese inclusive education teachers' agency within temporal-relational contexts. *Teaching and Teacher Education, 61*, 115–123.

World Health Organization/ World Bank. (2011). *World report on disability 2011*. Author.

CHAPTER 3

The Indian Context

The previous chapter examined the influence of neoliberal globalization and international development that promoted English as a global *lingua franca* and transposed universal educational structures and standards on a postcolonial "exclusionary school system, which is governed by a highly flawed selection mechanism..., not unlike those the British colonists established... that continue to haunt these (developing) societies even to this day" (Brissett & Bailey, 2021, p. 380). Moving from the macrocosm to the meso, this chapter analyzes the impact of globalization and development in India, which has seen tremendous upheaval since the 1990s with new values of rights and meritocracies jostling with age-old social hierarchies (Bhatia, 2018; Roy, 2019). It explores two aspects of the Indian social and economic context, class/caste and language, that have profoundly influenced access to education for children from lower castes/classes and multilingual backgrounds. It also examines the intersection of these factors with policies for education for all and language education and with the educational system.

The Indian Social and Economic Context

Independence in 1947 accelerated change and growth in India. In the 1990s, major neoliberal economic reforms propelled further growth, improving the lives of many overall. The "new India" of individual social

M. Kalyanpur, *Development, Education and Learning Disability in India*, Palgrave Studies in Disability and International Development,
https://doi.org/10.1007/978-3-030-83989-5_3

mobility generated by these changes enabled the millennial generation growing up in "mofussil" or smaller towns to compete with the traditional elite in previously unimaginable ways, not least by accessing English and thereby higher education, and amassing wealth and political power (Bhatia, 2018; Deb, 2011). Private wealth practically doubled between 2008 and 2018 (AfrAsia Bank, 2020), and the number of *billionaires* increased from 9 to 119 between 2000 and 2019 (World Economic Forum, 2019). By 2027, the billionaire count is expected to triple to 357 ("By 2027," Times of India, July 12, 2018).

Yet, 27.9% of India's population was living in poverty in 2016, according to the 2019 global Multidimensional Poverty Index; it had been at 55.1% in 2006. The Gini wealth coefficient, which measures wealth inequality with 100% denoting perfect inequality, was at 85.4% in 2019 (Chakravarty, 2020). Although in 2018, 66% of the population continued to live in rural areas, urban areas, particularly metropolises like Mumbai, were crowded, and 24% of the urban population lived in slums (World Bank, 2020). These wealth inequalities contribute to a sense of two Indias: the "beautiful forevers" of affluent lifestyles and the India behind these "beautiful forevers" (Boo, 2012), each parallel universe living cheek by jowl and dependent on the other for its existence (Bhatia, 2018; Deb, 2011). During my stay in India, I experienced this uneasy juxtaposition several times. One weekend, I was at the one-room home of a parent whose child attended a low-fee-paying private school. The next, I was at a cousin's country farmhouse with its cathedral ceiling and swimming pool. One week, I was in a school in a *basti* where 60 students sat crammed three to a bench in one classroom. The next week, I was in a school in the elite section of town where 30 students were in a brightly lit, airy classroom.

Caste, which in India is strongly linked with class, is a major social factor that intersects with these economic inequities (Loomba, 2016; Roy, 2019; Wilkerson, 2020; Yengde, 2019). The caste system, a predetermined belonging based on kinship groups and occupation, is deeply embedded in the hierarchical structure of Indian society (Béteille, 2012; Ghurye, 1936, as cited in Celarent, 2011; Loomba, 2016; Stalin, 2007). Although a systemic form of discrimination peculiar to India, the political and social elite have consistently sought to treat it as an internal cultural matter requiring Hindu religious reform rather than an infringement of sociopolitical rights (Loomba, 2016; Mosse, 2018; Yengde, 2019). Because of the historical correlation between caste and occupation, despite government reservation policies or affirmative action, the lowest castes or *Dalits* continue to face the discrimination of a minority status and be over-represented among the poor, embodying all the social indicators of poverty, including little or no access to education, health care, nutritional food or semi-skilled or skilled jobs (Loomba, 2016; Mosse, 2018; Yengde, 2019). Little change in societal attitudes have also perpetuated exclusionary practices, including where to live, exchange of food and water and access to places of worship (Beteille, 2012; Roy, 2019; Stalin, 2007; Yengde, 2019).[1] Bhatia (2018) noted:

> Class status in India is not just a reflection of one's socio-economic standing but also deeply connected to practices of consumption, marriage, sexuality, family values, education, work and lifestyle. Class identities are constructed and formed by their interaction with other cultural categories of caste, religion, and gender… Class in the contemporary Indian urban context is essentially rooted in social and economic hierarchies, but it also signifies politics of inclusion and exclusion, and it creates a social imagination in which some people are viewed as central and others are framed as outsiders in the globalization process. (p. xxx)

India is also a country of many languages. Although Hindi and English are the official languages, there are 22 official regional or state and 122 main languages ("Eighth Schedule," 2020). Most Indian children grow up multilingual, learning their mother tongue at home, often different from the regional language, and the regional and official language at school. English was intended to be phased out as the official language. However, with higher education being almost entirely in English and the increasing perception of English acquisition as the path to success, English has persisted (Mohan, 2017) and become increasingly a source of social disparity (Chattopadhay & Roy, 2017; Vulli, 2014). As Bhatia (2018) asserts, "the English-speaking class is the new caste" (p. 192). Yet, in 2019, only 15% spoke English (Ministry of Human Resource Development, 2019), almost all beneficiaries of private schools. The intersection of caste and language with wealth inequality has had significant implications for educational access and outcomes. The next section explores the issue of caste.

Caste and Class

Hinduism[2] is the oldest world religion and maintains its Brahmanic hegemony till today. India is the largest Hindu majority country with almost 80% of its population being Hindu, and about 95% of the world's Hindus living in India ("Hinduism," 2019).[3] It is unclear when the caste system, the most deeply embedded scourge of Hinduism, began, an issue which has created a political divisiveness within the current context of *Hindutva* or Hindu fundamentalism (Joseph, 2019; Shahane, 2019). Historian Singh (2009) suggested its origins may lie in proto-history while seminal sociologist G.S. Ghurye (1936, as cited in Celarent, 2011) concluded that caste was fundamentally the product of underlying racial differences, instituted to maintain ritual and racial purity between the Indo-Aryan outsiders and the indigenous Dravidians (Joseph, 2019; Shahane, 2019). The caste system is not easily explained. Essentially, there are four main castes: *Brahmins*, or the caste of priests, are the smallest and the most elite group, controlling access to sacred learning, and by default, political and economic power; *Kshatriyas*, or the caste of warriors and rulers; *Vaishyas*, or the caste of traders; and the two largest groups, the *Sudras*, the caste of field laborers, and the *Dalits*,[4] who performed so-called unclean tasks like washing clothes. The more ritualistic purification a caste and a sub-caste within this caste maintained, the higher its status.[5] When "intercultural contact, geographic mobility, occupational drift, religious change, and—above all—miscegenation" blurred these lines, "the principles of hierarchy, purity, and endogamy (were) redeployed to sharpen the boundaries again" (Celarent, 2011, p. 1716). During the late nineteenth century, the British who "sought to pigeonhole individuals into categories that were not always clear-cut or well-defined" (Béteille, 2012, p. 128) made the enumeration of caste—along with tribe, sect, religion and language—an established part of the Census practice (Samarendra, 2011). Post-Independence government reservation policies perpetuate caste status through school registries and personnel files in government offices. Loomba (2016) notes:

> Despite untouchability being constitutionally abolished in 1950, caste oppression is pervasive today. Over 160 million Untouchables—or Dalits—are subject to different forms of discrimination: they are denied access to places of worship, clean water, housing, and land; their children are still kept out of, or ill-treated within, schools. (p. 220)

In contemporary Indian society, caste difference is much less of an obvious external marker, although some Brahmins still choose to wear their high-caste markings on their foreheads. Indian surnames are a signifier, not only of one's caste, but also of one's region of origin and community; family names are a mine of information to the Indian insider, without having to ask anything more than "Your good name, sir?" Many *Dalits* have chosen to drop or change their surname to make this information less easily available. Easier access to inexpensive, synthetic clothing has also reduced some of the external signifiers on a day-to-day basis, particularly in urban areas. Most significantly, opportunities for individuals and families to go into business have eroded some of these external caste barriers (Deb, 2011; Yengde, 2019). As Pawar (2009) notes in her autobiography,

> Today, because of education and job opportunities, many Dalits have made a transition from being below the poverty line to the middle class. A Dalit looks no different from any other person in society. This has blunted the edge of casteism or discrimination. (p. xi)

However, the traditional preponderance of lower-caste Indians among the poor still means that the majority of *Dalits* in India continue to be discriminated against (Loomba, 2016; Roy, 2019; Stalin, 2007; Wilkerson, 2020; Yengde, 2019), with the "sticky floor" effect affecting the poorest of the poor (Mosse, 2018). As Ambedkar (1936, as cited in Loomba, 2016) asserted, "the caste system is not merely a division of labour. *It is also a division of labourers*" (p. 223, emphasis in original).

Caste and Education

This section describes the impact caste has had on access to education. Through the ages, the types and levels of education were commensurate with caste. While Brahmans males were given access to sacred learning, the middle and lower castes learned the craft of royal leadership, and trade and business. *Dalits*, however, were deemed unworthy of receiving any education (Nambissan, 2009; Vulli, 2014). In school in India, my classmates and I learnt the story of Eklavya, a young man who eavesdropped on the archery training that was being given to five young princes, the Pandavas. Worried that Eklavya's skill might surpass that of Arjun, his favorite pupil, the *guru* asked Eklavya to sacrifice his right thumb to the *guru*. Eklavya did so, and we schoolchildren collectively sighed at this wonderful gesture of loyalty and obedience. It was only as an adult that I learnt that Eklavya was eavesdropping because he was a *Dalit*, disallowed by caste to learn the art of archery which was the preserve of the warrior castes and princes, and the sacrifice was deliberately targeted at his violation of this social code to ensure that he would never use a bow and arrow again.

Caste and Education during Colonialism

Initially, the British decided it was more judicious to maintain this opprobrium toward this "most despised section" of society in fear that "if our system of education took root among them..., we might find ourselves... hated and despised by the (upper) castes" (Elphinstone, 1830, as cited in Upadhyay, 2012). However, this policy changed somewhat with the Caste Disabilities Removal Act of 1850. In 1882, educational policy stated that "no boy be refused admission to a Government college or school merely on the ground of caste" although it stressed that this principle be applied with due caution (Hunter, 1882, as cited in Nambissan, 1996, p. 1013). Separate government or state-aided missionary schools were set up for *Dalits* to avoid their coming into contact with upper-caste children. When allowed into ordinary schools, they were segregated by being seated on the verandah (Nambissan, 2002). Missionaries were particularly interested in educating *Dalits* as an opportunity to proselytize (O'Hanlon, 1985, cited in Upadhyay, 2012). Yet, in 1923, for every 1000 primary education students, there were 119 "advanced Hindus" and 18 "backward classes" (Upadhyay, 2012). In 1923, only 1% of *Dalit* schoolchildren went beyond primary school (Nambissan, 1996, 2009). Students learnt "the barest minimum of reading and writing skills" in one-room schoolhouses where their teachers were Muslims or lower-caste Hindus. *Dalit* students who were forced to attend common schools "faced virulent caste prejudice in the classroom" (Nambissan, 1996, p. 1013).

Caste and Education in Contemporary India

Attitudes and habits die hard. As Pandey and Pant (2008) asserted, "centuries of social ostracism and educational deprivation have resulted in negative school outcomes and poor achievement." Even in the new millennium, *Dalit* students attending school with the hope of discarding traditional occupations are often required to engage in jobs such as cleaning out the school toilets (Stalin, 2007; Yengde, 2019).

The Right to Education Act 2009

The Indian government has made concerted efforts to improve educational access for all. The 2009 Right to Education (RTE) Act, for instance, which obliges governments to provide and ensure compulsory admission,[6] attendance and completion of free elementary education by every child from ages six to 14 years, specifies that children belonging to "weaker sections" and "disadvantaged groups" are "not discriminated against and prevented from pursuing and completing elementary education on any grounds" (p. 4). However, RTE requires that schools maintain records on each child that includes, in addition to basic information such as name age, date of birth and address, "whether the child belongs to the weaker section… (or) to a disadvantaged group" (p. 4), allowing teachers, as McDermott et al. (2011) noted, to "perpetuate exclusion by referring to IQ or disability labels in combination with caste and tribal status."

RTE also mandates 25% caste-based reservations or quotas in private schools[7] in Grade 1 through admission by lottery (Juneja, 2017; Vulli, 2014). States then reimburse schools as part of the public-private partnership plan. Any admissions testing is forbidden, including an interview of the child or parent for admission. However, dismayed that this would reduce academic standards, some private schools made having an email address a requirement of an application, thereby excluding technically disadvantaged parents from seeking admission. In other situations, schools falsified the distance between home and school for poor applicants to disqualify them from admission.

The next section examines the socio-economic inequities arising from language education policies and the educational system that emerged from colonial/postcolonial contexts.

Language and Class

Language is equally subject to hierarchy and one way in which empires have sought to establish themselves is through a dominant language. Having linguistic cultural capital (Bourdieu & Passeron, 1990), or access to this dominant language, is a means to political and social power (Canagarajah, 2006), and more so in multilingual societies like India.

Language Policies and Education in Pre-British India

Here we explore language education policies through three periods, pre-British India, colonial India, and independent India..The origins of this legacy of multilingualism are rooted in the social hierarchy of Brahmanical ideology, synthesized into what we now know as Hinduism, where Sanskrit, as the language for sacred learning and royal inscriptions, was reserved for the upper-most caste, while people of lower status spoke the *prakrit* languages (Singh, 2009).[8] During the Mughal Empire, the Persian-speaking rulers made efforts to be conciliatory by allowing Sanskrit in the early stages, introducing Persian as the court language only during its imperial peak (Chavan, 2013; Parker, 2010). The aristocracy and other elites adopted Persian, while Sanskrit continued to be the bastion of Brahmins. Schools and higher education institutions, the *pathshala* for Sanskritic learning and the *madrassa* for Arabic and Persian, were set up where scholars traditionally received stipends to study (Chavan, 2013).

Language Policies and Education During Colonialism

In its early days the British Empire, which followed the Mughal, pursued the same conciliatory language policy of permitting Sanskrit and Persian, to the extent that the curriculum at the training college for British junior civil servants at Fort William in Madras (now Chennai) included English and Persian. In 1772 in Bengal, Governor General Hastings, himself fluent in Persian, enacted a policy aimed at "creating an Orientalized service elite competent in Indian languages and responsive to Indian traditions" (Kopf, 1984, p. 117), which had two major outcomes.[9] One, it resulted in the translation of major Persian and Sanskritic classics into English and Bengali, breaking the Brahminical intellectual monopoly. Two, it created a small, elite group of Indian intellectuals who were being exposed to Western science, history politics and literature through Bengali textbooks (Kopf, 1984). Although the learning of languages was mutual between Orientalists and this growing intelligentsia, it focused on introducing Indians to western traditions, medicine and intellectual thought as being inherently superior to Indian epistemologies.

In 1856, the British consolidated their empire by bringing in English as a symbol of "the triumph of British imperialism" (Chavan, 2013, p. 69), embodied by the Macaulay Minute on Education. While middle-caste Hindus were the first group to learn English, growing into the role of brokers, mediators and interpreters to the British, gradually they were joined by Brahmins (Chavan, 2013). The Macaulay Minute is redolent of the colonial assumptions of western superiority. For instance, it dismissed the significance of indigenous Indian languages and their contribution to academia and ideas as "contain(ing) neither literary nor scientific information" and concluded that "the superiority of the Europeans (was) absolutely immeasurable." The civilizing mission of colonialism of having "to educate a people who cannot at present be educated by means of their mother-tongue" provided the rationale for introducing "the English tongue (as being) the most useful to our native subjects." Colleges, and eventually universities, were established on the same lines as Oxford and Cambridge in England, with English as the only medium of instruction (Chavan, 2013). This resulted in a reconfiguration of elites into a new group of English-speaking intellectuals and civil administrators (Kopf, 1984).

Post-Independence Language Policies and Education

At Independence, with the Indian government configuring the states along linguistic lines, the *prakrit* languages, which historically had been accessible to the middle and lower castes, morphed into regional languages, and multilingualism flourished (Khubchandani, 1975). However, the governmental decision to maintain English as an official language, along with Hindi, spoken by a majority of people in northern India, meant that English speakers had the linguistic cultural capital that made them the elite. Although the intention was to phase out English, this never happened. Knowing English continued to be a marker of social status and a desideratum for most jobs, becoming even more desirable since neoliberal economic policies expanded the reach of contemporary globalization and reduced the poly-centrism of markets (Chattopadhay & Roy, 2017; Pritchett, 2013; Stromquist & Monkman, 2015). English speakers have easier access to the best higher education institutions nationally, where instruction is in English, or employment in the Indian administrative services, whose competitive entrance examination is conducted in English, and globally with transnational organizations (Mohan, 2017; Pritchett, 2013; Velaskar, 2015). Here, we consider two policies, the Three-Language Formula policy and the 2020 National Education Policy, that have sought, and failed, to reduce some of these inequities of access.

The Three-Language Formula Language Education Policy

At the national level, a complicated and controversial Three-Language Formula policy[10] was developed whereby one language was assigned the first place for study during the primary and secondary stages of education and second and third languages at the lower and higher secondary stages (Khubchandani, 1975). Government-run schools at national and state levels were conducted in Hindi and the regional language respectively, while private schools, such as the missionary schools introduced during colonialism, taught in English. Thus, the dominance of English and the relative devaluation of indigenous languages became "tied up with the issues of language privileges, cultural prestige, and socio-economic mobility" (Khubchandani, 1975, p. 152). The expectation, according to Khubchandani (1975), was that a student completing lower secondary education would be fluent in three languages: "mother tongue and two non-native modern languages, broadly, Hindi as official medium and a link language for the majority of people for interstate communication, and English as associate official medium and a link language for higher education and for intellectual and international communication" (p. 153).[11] However, in 1975, Khubchandani noted that about 80 languages were being used as the medium of instruction at different stages of instruction across the states. As a result, in 1992, only 2% of the entire population had access to English, and in 2019, only 15% spoke English (National Education Policy [NEP], Government of India, 2019).

The National Education Policy

The 2020 National Education Policy (NEP) continues to endorse the Three-Language Formula, but with a much greater reduction in the influence of English. As the draft NEP (Government of India, 2019) pointed out:

> English is regularly used by the elite as a criterion to determine whether someone is 'educated', and perhaps most unfortunately of all, as a prerequisite for jobs - even in cases of jobs where knowledge of English is entirely irrelevant.... This attitude has kept the elite class and the jobs they control segregated from the economically weaker sections of society... who happen not to speak the language of the colonists and current elite. It has created an unnatural aspiration of parents for their children to concentrate on learning and speaking languages that are not their own. (Government of India, 2019, p. 82)

The NEP 2020 encourages teachers to "use a bilingual approach" and avail of "high-quality bilingual textbooks and teaching-learning materials for science and mathematics," so that "students are enabled to think and speak about science both in their home language and in English" (Government of India, 2020, p. 16). However, given the impracticality of implementing this system and ensuring sufficient numbers of teachers trained in the academic vocabulary of the regional languages of their students and the availability of high-quality bilingual books in a variety of language configurations, these attempts to upend the linguistic hegemony and class-based privileges of knowing English have been critiqued as motivated by the government's right-wing, nationalist leanings toward a fundamentalist Hindu nation (Pundir, September 8, 2020).

Perhaps more significantly, within the current context of globalization and the linguistic imperialism of English, these policies can do little to subvert the status quo. Upper-income students will continue to receive an education in English and with a secondary focus on a regional or national language, while low-income students will continue to receive an education in their regional language and struggle to acquire English at the same level of fluency (Vulli, 2014). As Chattopadhay and Roy (2017) note, "In multi-lingual India, English is both a dominant language of commerce and higher education, and the official language of government transactions. Consequently, across all social classes, English is valued as a language for employability and social mobility. This in turn significantly drives the demand for 'English Medium' private schools in India (p. 3)." The next section explores how the hierarchies of the prevailing educational system have contributed to these inequities.

Hierarchies and Inequities in the Educational System

Despite all the public and private sector efforts to provide schooling for all, there are not enough schools to accommodate the millions of school-age children in India. The resulting bottleneck has created an immensely competitive educational environment that the school-leaving examination as a gate-keeping benchmark only exacerbates (Nawani, 2015). The most prestigious national or centrally regulated school-leaving exams are perceived to best prepare students for the admissions exams to national higher education institutes, including the National Institutes of Excellence, where English is the medium of instruction. State-regulated Board exams are seen as being in a lower tier. Even though the government has increased the number of institutions, and private universities and colleges have begun to take up the slack, the immense population of students competing for these relatively few spots has made these high stakes examinations an immense hurdle. Students have been known to commit suicide rather than take exams they expected to fail in or report their failing results to their parents (Nawani, 2015).

Since Independence, the disparities between government-run and private schools have become starker. Private schools, initially established by missionaries, were considered prestigious, as they offered an education where English was the medium of instruction (Annamalai, 2004; Waldrop, 2015). This gave its students an automatic leg-up in the competition for higher education. Government-run schools, on the other hand, offered an education in the regional language, essentially limiting opportunities for greater social and economic mobility for poor children. This concatenation of hierarchy of languages, wealth inequality and insufficient schools to meet demand is the backdrop against which students in India receive an education.

Government Schools

Universal primary education in India was spearheaded by the government in the 1950s through regional language, government-run schools. With World Bank funding, the SSA/EFA program further expanded access in the 1990s. Although the two decades following saw more growth than had occurred in the first four decades after Independence, resulting in an exponential demand for teachers (Manjrekar, 2013), there are still many shortcomings. Poorly resourced, often lacking trained or any teachers, particularly in rural areas, and with large class sizes, government-run schools have fallen into disfavor (Mukhopadhyay & Sriprakash, 2011). The number of SSA teachers has been steadily decreasing because the responsibilities are so onerous—for instance, in rural areas, they can be required to register the births and deaths occurring in the village (Sriprakash, 2011)—that many prefer to earn less at a private school than work as an SSA teacher. Sometimes class sizes can go as high as 140 students, especially in major metropolises with large urban poor populations, as in Mumbai (BMC) schools. As a result, implementing inclusive education has been difficult. Various studies found teacher absenteeism to be rampant, with less than half of teachers being both present and engaged in teaching on any given school day (see Pritchett, 2013). The gradual disenchantment with government schools was furthered by its persistence with regional language-based education and the general perception that the quality of education was better in private schools.

Private Schools

Although considered better than government schools, the educational structure of private schools is also tiered. Kurrien (2005, as cited in Shenoy, in press) identified two types of private schools: (a) English-medium middle-high-cost private schools, where teachers are proficient in English, but students have varying levels of exposure to English in their environment, including a home or first language; and (b) English-medium low-cost private schools, where teachers have limited proficiency in English and students have limited exposure to English in their environment, but parents aspire toward upward mobility via English. Students at high-cost schools, often called "convent schools" because of their missionary origins, tend to be fluent English speakers, their parents having been educated in English as well (LaDousa, 2017). Long the bastion of the upper middle classes and elite, the primary attraction of these private missionary schools was a curriculum in Western epistemologies and in English (Kumar, 2017; Motha, 2014; Waldrop, 2015).

In the 1980s, a burgeoning middle class began to demand quality private English-medium schooling. Education became the new growth industry within the private sector. Alternative schools offered access to innovative curricula and school-leaving examinations and programs, such as the International Baccalaureate.[12] The international recognition and equivalence of these school-leaving diplomas allowed students to pursue higher education internationally, while the flexible programming and innovative content made them particularly attractive to elite and affluent students struggling within the less flexible and more competitive environments of the national and state board exams. The exclusivity of the clientele of these schools was ensured by their tuition fees (LaDousa, 2017).

The neoliberal economic reforms of the 1990s, the emergence of English as the global *lingua franca,* and increasing dissatisfaction with the quality of schooling in government schools have led to a third tier of private English-medium school. These are the low-fee-paying (LFP) schools catering to low-income students, many of whom might be first-generation students and almost all with minimal exposure to English. Based on the same immersion model as the earlier tiers of English-medium schools, these schools implement "English only" policies in classrooms, so that their students are being forced to learn in English at the same time that they are learning English. The complex nature of this learning process is analyzed in more detail in Chapter 5.

Private schools can be aided or unaided. Aided private schools receive funding from the government, which controls the recruitment of teachers and pays their salaries; in return, the schools must maintain their fees within a stipulated limit (Chattopadhay & Roy, 2017). Unaided private schools receive no government assistance and set their own fees. The three LFP schools in this study were aided, while the three middle-high-cost schools were not. All were managed by education trusts, some under the Corporate Social Responsibility Act, which required corporations to contribute 3% of their profits toward social sector charitable development (Chattopadhay & Roy, 2017; Juneja, 2017; LaDousa, 2017). Private unaided and aided schools enroll 27.5 and 5.5% of the total student population, respectively, while constituting 17 and 5% of total schools nationally (NUEPA, 2013, as cited in Chattopadhay & Roy, 2017).

Private Low-Fee-Paying Schools

Private sector inroads into schooling have been touted as fortuitous and beneficial to all, particularly the poor (Heyneman & Stern, 2014; Tooley, 2009). However, closer scrutiny of LFP schools has found what Chattopadhay and Roy (2017) call "emerging fault lines" in the LFP sector. They challenge the assertion of affordability, noting that just by being fee-based, LFP schools continue to preclude access to the poorest of the poor, catering largely to low-income populations with more stable incomes (Chattopadhay & Roy, 2017). Further, LFP schools sacrifice quality of education to offset costs by hiring poorly qualified teachers, providing minimal curricular resources and maintaining large class sizes (Chattopadhay & Roy, 2017; Sarangapani & Winch, 2010), exercising "predatory capitalism" by leveraging low-income families' aspirations to acquire English (Juneja, 2017; LaDousa, 2017, p. 17). As Chattopadhay and Roy (2017) state:

> Poor parents, aspiring to model the middle class behaviors for their children, become willing participants in the "English medium project" of LFP

> schools that presents a mirage and a false choice. In LFP schools, poor children may neither attain knowledge of English nor gain understanding of subject matters that are taught in poor quality English. (p. 9)

Seeking to meet the "quality education" mandate of the Sustainable Development Goals, the RTE Act has developed regulations that LFP schools must comply with or risk de-recognition and eventually closure. However, since these regulations involve reporting on inputs and process, such as amount of furniture in classrooms and basic infrastructure like sex-separate toilets, rather than on outcomes (Pritchett, 2013), Chattopadhay and Roy (2017) concluded that compliance could lead to higher operating costs and student fees, which would make LFP schools unaffordable. When English as a medium of instruction is thrown into the mix of subtractive factors, and considering that for most of the students at these schools neither conversational nor academic English is a part of their everyday lives, the quality of education they receive is further lowered (Juneja, 2017; LaDousa, 2017; Mohan, 2017).

Conclusion

Disparities in caste and class have led to disparities in access to education and, in particular, to the language of power. Despite best efforts to "demonetize" the cultural capital of English, a colonial legacy resulting from a deliberate decision to create an English-speaking class that could administer the lower ranks of the imperial service, language education policies have not been able to overturn this legacy. Instead, its Three-Language Formula policy has tended to maintain the privilege of the English-speaking elite. The educational system has separated into government and private schooling, with increasing numbers preferring private schools in the belief that they offer a better quality of education. Private schools, however, and English-medium schools in particular, have evolved into a highly nuanced tiered structure that directly influences the quality of education students receive. Middle-high-cost schools serve the elite who already have ready and fluent access to English or the middle-class who are aspiring to improve their social and economic mobility through increased access to English. Low-fee-paying schools, on the other hand, attract poor students who have little everyday access to English and place them in situations where they are expected to acquire academic level English fluency and comprehension. In the end, as Jhingran (2009) asserted, "the language deficit that is forced upon the child the day she/he joins school becomes a learning deficit that only grows each year (p. 268)." The next chapter examines the microcosm of schools to explore further how knowledge and fluency with English becomes a social separator between teachers and students in LFP schools, mirroring many of the social hierarchies prevalent within the larger social structure. Knowledge of English becomes the filter by which parents of LFP students are measured and found lacking, by which teachers establish and assert their superior status in schools to compensate for some of the indignities of having to teach in low-paid, low-valued positions.

Notes

1. In 2018, the President of India, the second *Dalit* to be elected to this position, and First Lady were not permitted into the sanctum sanctorum during their maiden visit to the high holy Jagannath Temple in Odisha. Commenting on the incident, a party leader ironically noted, "Until now, only common devotees were exploited by the (temple) servitors. Now it seems not even the President and his family were spared" ("In Puri, priests block President," *Times of India*, June 27, 2018, p. 9).
2. A caveat to this statement: The word "Hindu" was originally a geographical term used in ancient Persian scriptures to refer to the lands beyond the Indus river. In the medieval period, the term came to acquire a religious-cultural meaning. The English word "Hinduism" was first used in 1816 by Raja Ram Mohan Roy (Singh, 2009, p. 433).
3. It has survived threats from within, resulting in more egalitarian religions of Jainism, Buddhism and Sikhism, all of which began as revolts against the elitism and hierarchies of Brahmanism (Omvedt, 2003). Later, it survived threats from without, weathering the domination of two proselytizing modern religions, Islam and Christianity, during the political subjugation by the Mughal Empire and the British Raj, respectively, each of which spanned close to 200 years. Despite these threats, today, just over 14% of the population is Muslim and just over 2% Christian. Sikhs also make up less than 2%, Buddhists constitute less than 1%, Jains less than half of one percent.

4. The term *Dalit*, which literally means "ground down" or oppressed, is a recent and self-selected descriptor (Pawar, 2009). During British rule, the term used was Untouchables, which was changed with Independence to the term Backward Castes used in official situations.
5. In Dutt's (2019) words, "the descriptors 'upper' and 'lower' to refer to people who belong to various caste groups are arbitrary, illogical and offensive. I use these terms since they are widely understood, but I reject the implied inherent 'superiority' or 'inferiority' that they have come to stand for" (p. x).
6. It clarifies that "compulsory" does not mean it is mandated that parents send their children to school but that it is compulsory for the government to provide free education to all children.
7. This was later modified to exclude private minority schools and boarding schools.
8. Although, in the south, the Dravidian languages originating from Tamil were, and have remained, distinct from the Indo-Aryan languages, they have been strongly influenced by Sanskritic and Brahmanic ideology (Singh, 2009). It is also situated in the geopolitical history of the region: Although the Indian sub-continent was called Hindustan to refer to the lands beyond the Indus River, it was far from the monolithic realm this suggests. Instead, it consisted of multiple tribal fiefdoms, kingdoms and empires, similar to those in Africa, which were constituted into a single nation state at the end of colonialism.
9. Known as British Orientalism and "designed to bridge the psycho-cultural gap between ruler and ruled (Kopf, 1984, p. 118)," this movement led to the establishment in Calcutta (now Kolkata) of the College of Fort William in 1800 and the Hindu College (which later became the prestigious Presidency College) in 1816.
10. Originally adopted in 1956, over the years, this policy has been adapted and since incorporated into the national education policy (Government of India, 2020).
11. Education being a state portfolio, tremendous variations have developed in terms of what this nomenclature signifies, such as differences between mother tongue, preferred language, home language, regional language, link language, national language, official language, or literary language. For instance, the 2011 Census found that many respondents differentiated between first language,

everyday tongue, most preferred language and mother tongue when asked to identify their primary language ("40%," 2018).

12. Other options include the International General Certificate of Secondary Education (IGCSE) exam administered by the University of Cambridge International Examinations. Typically, IB programs have a strong language component, which could have been exploited in a language-rich country like India; however, while students can choose to take an Indian language, most prefer to take a world language like Spanish or French.

References

AfrAsia Bank. (2020). *Global Wealth Migration Review 2019*. https://e.issuu.com/embed.html?u=newworldwealth&d=gwmr_summary_presentation_2019

Annamalai, E. (2004). Medium of power: The question of English in education in India. In J.W. Tollefson, & A.B. M. Tsui (Eds.), *Medium of instruction policies: Which agenda? Whose agenda?* (pp. 177–194). Routledge.

Béteille, A. (2012). *Caste and the citizen*. Oxford University Press.

Bhatia, S. (2018). *Decolonizing Psychology: Globalization, Social Justice, and Indian Youth Identities*. Oxford University Press.

Boo, K. (2012). *Behind the beautiful forevers: Life, death and hope in a Mumbai undercity*. Random House.

Bourdieu, P., & Passeron, J. C. (1990). *Reproduction in education, society and culture*. Sage.

Brissett, N. O. M., & Bailey, C. Y. (2021). Small Axe, episode 5, "Education" directed by Steve McQueen, produced by Anita Overland and Mike Elliott, and edited by Chris Dickens and Steve McQueen. *Comparative Education Review, 65*(2), 380–381.

By 2027, India will triple its billionaire count. (2018, July 12). *Times of India*, p. 7.

Canagarajah, S. (2006). Changing communicative needs, revised assessment objectives: Testing English as an international language. *Language Assessment Quarterly, 3*(3), 229–242.

Celarent, B. (2011). Caste and Race in India by G. S. Ghurye. *American Journal of Sociology, 116*(5), 1713–1719.

Chakravarty, M. (4 June, 2020). *Richest 10% of Indians own over 3/4th of wealth in India*. https://www.livemint.com/Money/iH2aBEUDpG06hM78diSSEJ/Richest-10-of-Indians-own-over-34th-of-wealth-in-India.html

Chattopadhay, T., & Roy, M. (2017). *Low-Fee Private Schools in India: The Emerging Fault Lines* (Working Paper 233). National Center for the Study of

Privatization in Education. Teachers College, Columbia University. http://ncspe.tc.columbia.edu/center-news/low-fee-private-schools-in-india/

Chavan, D. (2013). *Language politics under colonialism: Caste, class and language pedagogy in Western India*. Cambridge Scholars Publishing.

Deb, S. (2011). *The beautiful and the damned: A portrait of the new India*. Faber.

Dutt, Y. (2019). *Coming out as Dalit: A memoir*. Aleph.

Government of India. (2019). *Draft National Education Policy 2019*. Ministry of Human Resource Development.

Government of India. (2020). *National Education Policy 2020*. Ministry of Education.

Heyneman, S. P., & Stern, J. M. (2014). Low cost private schools for the poor: What public policy is appropriate? *International Journal of Educational Development, 35*, 3–15.

Jhingran, D. (2009). Hundreds of home languages in the country and many in most classrooms: Coping with diversity in primary education in India. In T. Skutnabb-Kangas, R. Phillipson, A. K. Mohanty, & M. Panda (Eds.), *Social justice through multilingual education* (pp. 263–281). Multilingual Matters.

Joseph, T. (2019, September 13). There's no confusion: The new reports clearly confirm 'Arya' migration into India. *The Hindu*. https://www.thehindu.com/society/history-and-culture/theres-no-confusion-the-new-reports-clearly-confirm-arya-migration-into-india/article29409611.ece

Juneja, N. (2017). Education in urban areas. In K. Kumar (Ed.), *Routledge handbook of education in India, debates, practices, and policies* (pp. 26–42). Routledge. https://www.routledgehandbooks.com/doi/10.4324/9781315107929.ch2

Khubchandani, L. M. (1975). Dilemmas of language transition: Challenges to language planning in India. *ERIC Document Reproduction Service No. ED, 116*, 474.

Kopf, D. (1984). Orientalism and the Indian educated elite. In G. P. Kelly & P. G. Altbach (Eds.), *Education and the colonial experience* (pp. 117–136). Transaction Books.

Kumar, N. (2017). The technology and politics of teaching English. In M. J. Tiwary, S. Kumar, & A. K. Mitra (Eds.), *Dynamics of inclusive classrooms: Social diversity, inequality and school education in India* (pp. 183–202). Orient BlackSwan.

LaDousa, C. (2017). Mind the (language-medium) gap. In K. Kumar (Ed.), *Routledge handbook of education in India, debates, practices, and policies* (pp. 81–96). Routledge. https://www.routledgehandbooks.com/doi/10.4324/9781315107929.ch5

Loomba, A. (2016). The everyday violence of caste. *College Literature, 43*(1), 220–225.

Manjrekar, N. (2013). Women school teachers in new times: Some preliminary reflections. *Indian Journal of Gender Studies, 20*(2), 335–356.
McDermott, R., Edgar, B., & Scarloss, B. (2011). Global norming. In A. Artiles, E. Kozleski, & F. Waitoller (Eds.), *Inclusive education in five continents* (pp. 223–235). Harvard Education Press.
Ministry of Human Resource Development. (2019). *Draft National Education Policy 2019*. Government of India.
Mohan, P. (2017). The road to English: The slow migration of the EWS child to elite India. In M. J. Tiwary, S. Kumar, & A. K. Mitra (Eds.), *Dynamics of inclusive classrooms: Social diversity, inequality and school education in India* (pp. 203–215). Orient BlackSwan.
Mosse, D. (2018). Caste and development: Contemporary perspectives on a structure of discrimination and advantage. *World Development, 110*, 422–436. https://doi.org/10.1016/j.worlddev.2018.06.003
Motha, S. (2014). *Race, empire, and English language teaching: Creating responsible and ethical anti-racist practice*. Teachers College Press.
Mukhopadhyay, R., & Sriprakash, A. (2011). Global frameworks, local contin gencies: Policy translations and education development in India. *Compare: A Journal of Comparative and International Education, 41*(3), 311–326.
Nambissan, G. B. (1996, April 20–27). Equity in education? Schooling of Dalit children in India. *Economic and Political Weekly*, 1011–1024.
Nambissan, G. B. (2009). *Exclusion and discrimination in schools: Experiences of dalit children*. Indian Institute of Dalit Studies and UNICEF.
Nambissan, G. B. (2012). Private schools for the poor: Business as usual? *Economic & Political Weekly, 47*(41), 51–58.
Nawani, D. (2015). Rethinking assessments in schools. *Economic & Political Weekly, 50*(3), 37–42.
Omvedt, G. (2003). *Buddhism in India: Challenging Brahmanism and caste*. Sage.
Pandey, J., & Pant, S. K. (2008). The marginalized Dalits: Disadvantages in learning. In K. Thapa & G. M. van der Aalsvoort (Eds.), *Perspectives on learning disabilities in India: Current practices and prospects*. Sage.
Parker, C. (2010). *Global interactions in the early modern age, 1400–1800*. University Press.
Pawar, U. (2009). *The weave of my life: A Dalit woman's memoir* (M. Pundit, Trans.). Columbia University Press.
Pritchett, L. (2013). *The rebirth of education: Schooling ain't learning*. Brookings Institution Press.
Pundir, P. (2020, September 8). *Why does India keep dealing with 'Language Wars'?* https://www.vice.com/en/article/4aymeb/why-does-india-keep-dea ling-with-language-wars

Roy, A. (2019). Foreword. In A. Roy (Ed.), *My seditious heart: Collected nonfiction* (pp. xiii–xxv).

Samarendra, P. (2011). Census in Colonial India and the Birth of Caste. *Economic and Political Weekly, 46*(33), 51–58.

Sarangapani, P. M., & Winch, C. (2010). Tooley, Dixon and Gomathi on private education in Hyderabad: A reply. *Oxford Review of Education, 36*(4), 499–515. https://doi.org/10.1080/03054985.2010.495465

Shahane, G. (2019, September 14). Why Hindutva supporters love to hate the discredited Aryan Invasion Theory. *Scroll.in*. https://scroll.in/article/937043/why-hindutva-supporters-love-to-hate-the-discredited-aryan-invasion-theory

Singh, U. (2009). *A history of ancient and early medieval India: From the Stone Age to the 12th century*. Pearson.

Stalin, K. (Director). (2007). *India Untouched: Stories of a People Apart* [Film]. Dhrishti Films. https://www.cultureunplugged.com/documentary/watch-online/play/5752/India-Untouched-https://www.documentarystorm.com/india-untouched/

Stromquist, N. P., & Monkman, K. (2015). Defining globalization and assessing its implications for knowledge and education, revisited. In N. P. Stromquist & K. Monkman (Eds.), *Globalization and education: Integration and contestation across cultures* (2nd ed., pp. 1–20). Rowman & Littlefield.

Sriprakash, A. (2011). Being a Teacher in Contexts of Change: Education Reform and the Repositioning of Teachers' Work in India. *Contemporary Education Dialogue, 8*(1), 5–31

Tooley, J. (2009). *The beautiful tree: A personal journey into how the world's poorest people are educating themselves*. Penguin.

Upadhyay, S. B. (2012). Dalits and the modern education in colonial India. In C. Sleeter, S. B. Upadhyay, A. K. Mishra, & S. Kumar (Eds.), *School education, pluralism and marginality: Comparative perspectives* (pp. 152–180). Orient Blackswan.

Velaskar, P. (2015). Educational stratification, dominant ideology, and the reproduction of disadvantage in India. In. M. Thapan (Ed.), *Education and society: Themes, perspectives and practices* (pp. 428–448). Oxford University Press.

Vulli, D. (2014). English and medium of instruction: Dalit discourse in Indian education. *Research Journal of Educational Sciences, 2*(2), 1–6.

Waldrop, A. (2015). The meaning of the old school tie: Private schools, admission procedures, and class segmentation. In. M. Thapan (Ed.), *Education and society: Themes, perspectives and practices* (pp. 405–427). Oxford University Press.

Wilkerson, I. (2020). *Caste: The origins of our discontents*. Random House.

World Bank. (2020). *India statistics*. https://tradingeconomics.com/india/rural-population-percent-of-total-population-wb-data.html

World Economic Forum. (2019). *India's mega-rich are on the rise*. https://www.weforum.org/agenda/2019/07/india-mega-rich-are-on-rise-inequality/
Yengde, S. (2019). *Caste matters*. Penguin Viking.

CHAPTER 4

Social Hierarchies in the School

The transformative consequences of contemporary globalization and international development with regard to Education For All (EFA) for millions of children cannot be understated. Close to 100% of primary school-aged children are now in school in India, including children who would have not traditionally received an education, like girls and children from low-income and Dalit backgrounds (Ajmera, 2016; Singal et al., 2019). As scholars note, even the poorest of the poor and the most socially disadvantaged have begun to recognize education as a means of social and economic mobility (Singal, et al., 2019; Wagner et al., 2018).

However, some of the consequences have been problematic. Globalization has also contributed to the postcolonial emergence of English as the global *lingua franca* (Annamalai, 2004; Vulli, 2014), and the structural adjustment policies of international development have reduced public sector expenses and encouraged private–public partnerships in education and health (Chabbott, 2015; Pritchett, 2013). This combination of events has led to an increasing demand for English, traditionally an elite reserve, by middle- and low-income families, resulting in the proliferation of private, low-fee-paying (LFP), English-medium schools (Day et al., 2014; Heyneman & Stern, 2014; Tooley, 2009). In 2017, the share of private schools in India was over 63% of all urban schools, with their enrollment share exceeding 70% of all urban enrollment (Juneja, 2017).

M. Kalyanpur, *Development, Education and Learning Disability in India*, Palgrave Studies in Disability and International Development,
https://doi.org/10.1007/978-3-030-83989-5_4

Unfortunately, market-driven measures to increase efficiency and productivity in LFP schools have led to less infrastructure, lower teacher salaries and increased class sizes (Bangay & Latham, 2013), affecting the quality of education.

Educational expansion also spiked teacher recruitment (Belliappa, 2014; Sriprakash, 2011). The emphasis of international aid agencies to incorporate gender mainstreaming strategies into national policy (Manjrekar, 2013) led to women being recruited into a traditionally male-dominated, highly regarded profession (Ganapathy-Coleman, 2014). As a result, women constitute around 65% of all teachers in urban areas (DISE, 2008–2009, as cited in Manjrekar, 2013). However, cost-cutting by lowering salaries has led to the profession's devaluation with less qualified women teachers working in spaces of low pay and insecure employment (Chattopadhay & Roy, 2017; Manjrekar, 2013; Sriprakash, 2011). Although private, English-medium, LFP schools have responded to low-income families' increasing willingness to pay for access to English, this market ideology severely compromises quality.

Further complicating the situation are the social upheavals of class/caste and English fluency that underlie these economic transformations. The wealth inequalities of India manifest in the educational system: For instance, the differences in infrastructure and class size between LFP schools and upper-tier schools are obvious, stark and tangible. However, there are also subtle, more insidious differences relating to perceptions of social class and hierarchies of English, such as "convent-school" and regional accents (Deb, 2011; LaDousa, 2017), that affect the quality of education adversely for LFP students and lead to very different learning outcomes for low-income students and for middle-class and affluent students (Vulli, 2014).

This chapter explores the consequences of these transformations at the macro- and meso-levels as they play out at the micro-levels of schools and affect the lives of teachers and students. First, it examines these overt differences of infrastructure and the gendering of teaching. Next, it explores the less obvious differences of language and caste through two themes: teachers' own facility with English and their attitudes toward their students and their families as manifested through their perceptions and interactions. Situating the LFP schools and the teachers in this study within the macro and meso sociocultural contexts of globalization and national policies acknowledges their limited agentic control, or the power or opportunity to make choices or exert control over the system

(Bandura, 1999; Belliappa, 2014; Manjrekar, 2013), and recognizes that their attitudes and behavior are reflective of their realities, their social worlds and their own aspirations. Exploited themselves through underpay and overwork, their work is a means to an end: the aspiration to a better lifestyle for their own children (Sriprakash, 2011), just like the parents of their students.

School Infrastructure and Its Effects on Quality of Education

A main factor that differentiates school status is the social class of the students attending the schools, which, to some extent, is determined by the range of tuition charged (LaDousa, 2017). To maintain their status as LFP schools, school fees are capped (Chattopadhay & Roy, 2017). There are no restrictions for upper-tier schools. Tuition, then, becomes a gatekeeping mechanism and allows schools to establish a level of infrastructure and qualified teachers to target middle class and affluent clientele (CES, 2014). This section compares the infrastructure in the LFP schools with the upper-tier schools.

Scholars assert the lack of a global metric for what is meant by "quality" education has compelled individual governments to define it (Datt & Wang, 2020) and that India, for instance, has chosen to focus on inputs and processes rather than actual learning (Pritchett, 2013). The 2009 Right to Education Act provides standards on furniture in classrooms and basic infrastructure such as number of classrooms, separate toilets for boys and girls, drinking water facility, kitchen for mid-day meals and playground size (Babu, 2018; Chattopadhay & Roy, 2017; Ministry of Law & Justice, 2009). Standards on teacher capacity and teaching–learning inputs, including teacher qualification, minimum teacher salary and teacher-to-student ratio, are left to state and local governments (Chattopadhay & Roy, 2017). The notion of quality is further compromised by the market-driven focus on efficiency and productivity of neoliberal education sector reforms. Expressing "concerns for equality and equity of educational opportunity, especially in cities where private schools are concentrated" (Juneja, 2017, p. 29), scholars maintain most LFP schools sacrifice quality of education to offset costs by maintaining large class sizes, hiring poorly qualified teachers and providing minimal curricular resources (Chattopadhay & Roy, 2017; Sarangapani & Winch,

2010). For instance, describing the Gyan Shala model of LFP schools, Bangay and Latham (2013) noted,

> costs are kept low primarily due the fact that the classrooms are rented, single rooms and there are no playgrounds or other amenities and the teachers are hired from the informal sector at a fifth or a sixth of the salaries of tenured government teachers. (p. 249)

Infrastructure in LFP Schools

Students in the LFP schools in this study were housed in three-storied, concrete buildings built for the purpose, surrounded by tall brick walls, which also enclosed mud-packed playgrounds or a courtyard, spaces that were a luxury in a city where land was at a premium. In the morning, the buildings were a primary school for Grades 1 to 5. Each grade had three to five sections, with about 50 students in each section. In the afternoon, the buildings became secondary schools for Grades 6 to 10. In two schools, a small group of preschoolers also attended in the morning. All the schools had toilets with running water, separated by sex, that were regularly cleaned by custodial staff. Electrically operated water coolers for drinking water were available on every other floor. As part of the accessibility features for SSA or inclusive education, one school had a ramp at one entrance and an elevator. Two schools had computer labs consisting of two rows of computers on tables in locked rooms. According to one school principal, an NGO had donated the computers, but because no maintenance service was provided with the equipment, the room was kept locked. Although these resources may seem sparse, the fact that these schools had buildings for the purpose, playgrounds and even a computer room points to more than some LFP schools have (Bangay & Latham, 2013; Bhattacharya, 2013).

Indeed, many LFP schools struggle to meet the national standards for infrastructure (Babu, 2018; Bhattacharya, 2013). Here too, the small-sized classrooms were often crowded with furniture taking up the entire space in the room. In one school, the front row of desks abutted the teacher's platform at the front of the classroom. In another school, the desks were quite new and combined wood and metal frameworks; in the other two schools, the all-wooden desks were older. The desks were narrow, which resulted in books or pencil boxes falling off them frequently. On average, two students shared a desk comfortably; however, in many classrooms, three students sat at a desk because there were not enough desks for the number of students. Classroom arrangements such as these have implications for child-centered instruction, as options like placing students in small groups, incorporating movement-based activities or accessing the back of the room become harder to execute (Bhattacharya, 2013). As the National Curriculum Framework (NCERT, 2005) pointed out:

> Many schools invest in heavy metal benches and long desks, which can only be placed in rows, and which reinforce the teacher and blackboard-centred system of learning. Worse still, many of these do not have adequate place for children to keep their books and belongings, nor are they wide enough or with back support suitable for the physical comfort of the child. (p. 80)

As in most LFP schools (Bhattacharya, 2013; Boruah, 2017; Chattopadhay & Roy, 2017), the primary curricular resource the teachers had access to were textbooks. The teachers in one school had undergone professional development on using manipulatives or "play ways." However, the school had not invested in these resources, and the teachers had sought the parents' support in sending low-cost materials, such as number and letter cards made from "tailor board," to augment their instruction.

Infrastructure in Upper-Tier Schools

In contrast, the upper-tier schools were in more middle-class neighborhoods and fell into high or middle status by the range of tuition charged (LaDousa, 2017; Shenoy & Wagner, 2019). One school was a top status school, often referred to as a "convent school" as part of the legacy of missionary schools from colonial rule (Juneja, 2017; Waldrop, 2015). A second school was a middle-status school established in the last decade because of the increase in private sector involvement in education (LaDousa, 2017). The third school was a special school for students with learning disabilities, its fee structure placing it within the top status bracket. Stepping into these schools was almost like entering a school in the US. The buildings had bright, clean, airy spaces, colorful artwork and quality workmanship. The meeting rooms were air-conditioned, and the classrooms had smartboards. There were designated spaces to eat so students did not have to eat their lunches and snacks at their desks as in the LFP schools. All three schools had covered play areas, where students could run around during their recess and lunch and have their physical training/education (PE/PT) classes even during the monsoon season. Class sizes ranged between 30 and 40, and even fewer in the special school. The classrooms were large enough that teachers were able to be more creative with the space and move furniture as needed. Struggling students received pull-out supplemental services in small groups in a learning resource center, or resource room, for which parents paid an additional fee in the general education schools. The special school also offered counseling and reading specialist services, occupational, physical and speech therapy, training in information and computer technology (ICT), drama, gym/PE and art.

Given that a primary difference between the LFP and the upper-tier schools is access to more revenues, comparing their resources might be stating the obvious. The purpose of highlighting these differences, however, is to question the value of an education that does not guarantee its promised outcomes of quality. Even within a context of EFA, the comparatively resource-rich educational environments that are afforded to middle- and upper-class students underscore the limited choices that low-income families have, resulting in a lower quality of education for their children. It has been argued that private sector inroads into schooling for the poor are justified because they fill a gap left by governments and provide access to an education many millions would otherwise have been denied (Bangay & Latham, 2013; Day et al., 2014; Tooley, 2009). However, the neoliberal argument that the private sector's focus on cost-cutting and productivity is appropriate because it makes education for the poor available, with the embedded message that poor families should be grateful for any opportunity to an education, regardless of quality, is problematic.

The complexity of globalization and international development has also gendered the profession of teaching, resulting in low-paid and less valued positions as more women enter the field. The next section examines this more closely.

Teaching as Gendered Labor

It was not a coincidence all the teachers I interviewed were women. For one, women teachers "dominate recruitment in private LFP schools, working on insecure contracts, with very low pay and unregulated hours of work" (Manjrekar, 2013, p. 349). Additionally, teaching is gendered labor (Manjrekar, 2013). Considered a respectable profession for women because the work hours are compatible with family life and particularly with bringing up children (Belliappa, 2014; Manjrekar, 2013; Sriprakash, 2011), "the spaces of education are presumed to be feminised spaces, (and) teaching is a natural extension of their mothering capacities" (Manjrekar, 2013, p. 348). While all the LFP teachers in this study said they enjoyed teaching, they also mentioned that it was a convenient profession, because it allowed them to "do the housework" and earn a supplemental income that would help them to achieve their aspirational lifestyle. Having only completed a diploma or a certificate in education, a lower tier in teacher training than a bachelor's degree, the LFP teachers could not get hired in a middle-status private school or a government school where the pay would have been higher. Instead, as Sriprakash's study (2011) corroborated, their earnings contributed to family income, thus enabling them to send their own children to what they described as "better" schools, typically middle-status schools.

In this study, teachers and administrators alike believed family came first—that the single women would leave when they got married, the married women would do the same when they started their families, and the mothers would take leave in a family crisis. As an administrator who started as a teacher said in an interview:

> I didn't want to become a teacher. I told my mother, you just come home with corrections all the time! But my mother also persuaded me, see, from my own example, you can see how teaching is better suited for you, so you can have a family also. And now I see that she and my father were right.

Similarly, during an interview with a male Board of Trustees member, a teacher came around with *laddoo*, a sweetmeat, to celebrate her engagement, provoking the wry comment from him:

> This is why I prefer to hire married ladies. She will join her husband and leave (the school).

Although there are no statistics on teacher attrition in India (Menon, 2019), this prioritizing contributed to high turnover. Two women principals mentioned in interviews that turnover was always a concern because of "marriage and childcare." During this study, two teachers left to get married and one left after she had her baby. Several teachers who were married said they had chosen to work at this school because it was close to their home once they had moved to live with their husband's family. As one of them explained in an interview:

> I was teaching at X school in (a neighboring suburb) for two years. But when I got married, X school was too far. My husband's family didn't want me to travel. Also, I was now having family responsibilities. So I moved to this school.

Several teachers said they had taken a break while their children were very young since they did not have any family help with childcare and had returned to work now that their children were older. During a 25-minute observation, one teacher received three phone calls, explaining, "It's my mother calling. My child is very small and at home with my mother." Another teacher described in an informal conversation how her day began at four in the morning when she cooked hot packed lunches for her husband, school-going child, and herself and a fresh lunch for her in-laws, who looked after her preschooler while she was herself at school, and then came home and cooked dinner for the whole family, while supervising her son's homework. She said, "At the end of the day, I clean the whole house and then go to sleep." As a result of these familial pressures, teachers were frequently absent. Lacking a system of substitute teachers, an absent teacher meant the students in that class were distributed among two sections, further increasing class sizes and affecting continuity of instruction for the students.

These teachers are products of the larger, market-driven ideology of low-pay jobs and limited expectations for professional expertise on the one hand, and the social confines of respectable professions for women and familial expectations on the other. They are also products of aspirational lifestyles themselves, their incomes contributing toward their children's educational advancement. Like the teachers in Sriprakash's study (2011), they all lived in localities outside their schools, some requiring long commutes. As Arvind (2015) noted, this spatial distancing leaves no time to get to know the school neighborhood and community, resulting in a social distance between teachers and parents and affecting teachers' attitudes toward their students and their families. The next section examines the effects of social inequalities on the quality of schools. Specifically, the dynamics of traditional caste-based attitudes and "English-as-the-new-caste" (Bhatia, 2018) on the quality of education are explored through two themes: the teachers' own facility with English and their attitudes toward their students and families through their perceptions and interactions with them.

Spoken English as an Indicator of Social Class

Global and postcolonial factors have contributed to a hegemony of English as a universal *lingua franca* that privileges its acquisition as a highly desirable skill for a world-wide labor force (Annamalai, 2004; Vulli, 2014). Private, LFP and English-medium schools prey on families' aspirations for social and economic mobility through the acquisition of English (Bhattacharya, 2013; Chattopadhay & Roy, 2017; Sriprakash, 2011). The strong overlap of caste distinctions with class in India adds an additional layer of complexity to the situation, further exacerbated by contemporary rightwing extremist politics resulting in the more explicit othering of Dalits and Muslims (Guha, 2020; Sathyamala, 2019; Yengde, 2019). This section examines how these influences are manifested in the LFP schools through the teachers' own facility with spoken English and their perceptions of and interactions with their students' families.

Why English Medium?

While the British Raj severely and deliberately limited access to English to the elite classes as part of their "downward filtration" policy (Upadhyay, 2012), this stratification continued well past independence when states were established along linguistic lines. As a result, even as recently as 2019, only 15% of the population spoke English (Draft National Education Policy, 2019).[1] However, the increasing global acceptance of English as the universal *lingua franca* has compelled low-income parents to seek English-medium private education for their children (ASER, 2019; Kumar, 2017; Mohan, 2017). The aspirational lifestyle exemplified by the acquisition of English as a pathway to socioeconomic mobility even led to the creation of a Dalit goddess of English, who if prayed to, would give Dalits "the privilege of English to get respect and jobs" (Pandey, 2011). Bourdieu's theory of linguistic cultural capital and unexamined exclusion suggests that a "pre-established harmony" between historical legacies and educational systems often results in a "social selection" which legitimizes cultural reproduction and converts social hierarchies to academic hierarchies, thus perpetuating social inequalities as "teachers' judgements on their pupils transmute social classifications into school classifications" (Bourdieu & Passeron, 1990, p. xxiv). Within the Indian context, the perceived and actual social and economic advantages of English-speaking elites, as part of the postcolonial historical legacy, presented in Chapter 3, have provided the justification for English-medium instruction (EMI) in schools, thus re-creating social class structures in schools. Since the "new India" of the 1990s (Deb, 2011), the proliferation of LFP schools that offer EMI has attracted families from these aspirational strata.

Use of English in Schools

Private schools, particularly English-medium schools, are tiered by their tuition fees which demarcate the socioeconomic status of their students and teachers. Students from affluent backgrounds who can afford to attend top-tier schools are part of the Indian, English-speaking elite, as are their teachers. Students who attend low-fee-paying schools are primarily first-generation students and English speakers and tend to be taught by teachers who are themselves first-generation English speakers.

LFP Teachers' Facility with English

Teachers at the LFP schools had learned English at school, rather than at home. Although interviews were conducted in the hybridized "Indian English"[2] with me, among themselves and in the teachers' lounge, they spoke in Marathi or Hindi. One teacher mentioned her primary home language was Marathi; however, her own children now preferred speaking in English at home, and she was forced to speak to them in English. These teachers had been hired at these LFP schools because of their English-speaking skills. However, as second or third language English speakers, they engaged in "complex language encounters, a classroom context in which 'teachers and their learners engage with each other in a language which neither party can use with ease" (Evans & Cleghorn, 2010, p. 32, as cited in Bhattacharya, 2013). As teachers, they were constrained both by the fact that English was not their first language and that they had not received any instruction in how to teach English as a second language (Anderson & Lightfoot, 2018; Bhattacharya, 2013; Chattopadhay & Roy, 2017).

Further, the vocabulary of their Indian English was often quite different from the academic English used in school textbooks. During an observation I conducted, for instance, when a teacher solicited the names of classroom items from the students, she offered the Indian English words they were more familiar with, like "rubber" for eraser and "color" for crayon. However, when writing the words on the blackboard, she used the unfamiliar vocabulary words from their textbook "to learn proper words nicely," as she put it. As Bhattacharya (2013) notes, in adhering to academic English during formal instruction, the teachers made English even more inaccessible to their students.

Upper-Tier Teachers' Facility with English
On the other hand, English is the cornerstone of hierarchy in upper-tier private schools (Waldrop, 2015). The teachers at the upper-tier schools came from backgrounds where English was spoken at home. The teachers at the high status and special schools spoke the English of elite India, often called "convent school English" (Arvind, 2015; LaDousa, 2017), which is fluent and unaccented by any regional language undertones (Deb, 2011). The teachers at the middle-status school had attended similarly middle-status schools themselves and were now fluent and comfortable enough in English to teach in it; having made the leap from middle to upper-middle class, they now spoke English at home as well with their children. Their own fluency in English rendered moot the complication of medium of instruction in a language unfamiliar to them, as occurred in the LFP schools (Bhattacharya, 2013), which made teaching curricular content easier, providing opportunities for more creative and enriched instruction.

Language Skills of Students and Families

Almost all Indians are bilingual, if not multilingual ("Hindi mother tongue of 40% of Indians," June 27, 2018; Khubchandani, 1975; Mohan, 2017). They speak a home language, which may or may not be the same as the regional or state language. If different, they are likely to speak the regional language as well. They may also be familiar with Hindi, an official language, perhaps as a subject in school, through media exposure of television and Bollywood, or if they live in an urban area. However, their access and, therefore, their fluency in English depend on their socioeconomic status (Vulli, 2014).

LFP Students' and Families' Language Skills

All the LFP students came from the same low socioeconomic neighborhood in which their schools were located. Some were deemed so poor they received state-supplied free uniforms and fee subsidies to attend school. As is typical of LFP schools (Bhattacharya, 2013; Boruah, 2017; Mukhopadhyay, 2020), all students were introduced to both conversational and academic English for the first time in school. While most students spoke Hindi, Marathi and/or Urdu at home, a few also spoke Bhojpuri or Tamil. Their parents also spoke no English; interviews with them were conducted in Hindustani. Many, particularly the migrant families, had received no or little education. One mother worked as a maid; the other mothers said they had no source of income other than their husband's. The fathers worked as construction laborers, rickshaw drivers and street vendors. Only one father owned a scooter and co-owned a retail shop.

To be considered low fee, a government mandate required schools to keep monthly tuition fees below \$10, around Rs.600 (Chattopadhay & Roy, 2017). The LFP schools in the study charged between \$6.40 (Rs.450) and \$9.60 (Rs.550). Although the teachers and school management stressed the low fee, the parents mentioned this cost as a hardship, especially as there were additional expenses like schoolbooks and uniforms. However, the most significant extra expense was tuition classes, which cost as much as the school tuition, discussed in more detail in Chapter 6. These latter services were a parallel educational structure that had emerged to respond to the inadequacies of both the government and private school systems, providing additional, more personalized support that the large class sizes could not (Chattopadhay & Roy, 2017; Sarangapani & Winch, 2010).

The parents iterated the importance of their children learning English. The move to the city was for the next generation that would be educated, know English, earn well and become middle-class. Bhatia's (2018) description of *basti* residents in Pune, a city close to Mumbai, is appropriate for the families in this study: "Their aspirations are shaped by their caste identities, structural conditions of poverty, their schooling in the vernacular language, collective social practices, and the prestige accorded to speakers of English in urban India" (p. 166).

All the low-income parents in this study acknowledged it was imperative their children learn English. In an interview, a parent said:

> Everything now needs knowledge of English. Any job, anything, you can't progress without English. Okay, I didn't learn English but at least my children can and should. That's how we progress.

Similarly, one father, who had just moved from being a joint to a nuclear family, explained in an interview why he had placed both children in the English-medium school:

> My daughter has learnt English. He too will learn. His teacher has told me he is a little weak in school, but he's young, he'll pick up slowly. My wife and I can't help because we only speak Hindi. Now that we have moved into our own home, we will be able to put him in private tuition. It is a problem that we speak Hindi at home, and the school is English-medium. But we don't have a choice. If we want our children to do well and earn well, they must learn English.

Having been deprived of wealth, status and education for generations because of their caste/class (Vulli, 2014; Yengde, 2019), to the parents, access to English through EMI was now a primary means of socioeconomic mobility for their children.

Upper-Tier School Students' Language Skills

While there were no scholarship students at the two general education schools, the special school for students with learning disabilities did have a few. The students at the top status school came from English-speaking backgrounds (Kurrien, 2005, as cited in Shenoy & Wagner, 2019). They were children of the higher echelons of society, conversant with American pop culture. They owned cars, vacationed abroad and in India, lived in luxury apartments or homes in gated communities and consumed western goods (Bhatia, 2018; Boo, 2012; Deb, 2011). At school, they spoke fluent, "unaccented" (Deb, 2011) English among themselves in informal conversations. Students in the middle-status school were middle class. Their families owned a family car, went on vacations within the country, had cable television which gave them access to all the latest American shows and movies and enjoyed Bollywood movies as well. Although they spoke English well, their accents were "regional" and therefore less prestigious (Deb, 2011). In addition, they often spoke among themselves in Hindi or Marathi, especially during quick, urgent conversations in class (Kurrien, 2005, as cited in Shenoy & Wagner, 2019). These students are like those in LaDousa's study (2017) who could identify no benefits to learning Hindi but listed several for knowing English.

Many students at the special school were former students at the other two schools, having withdrawn from them to avail of the customized services the special school offered. Most came from English-speaking backgrounds and spoke English fluently; the scholarship students and some of the students who needed more curricular adaptations spoke Hindi, Gujarati or Marathi more fluently. As the teachers pointed out, although the school was an English-medium school, individualization seemed to dictate the need for teaching in the student's preferred language. As a result, many chose to provide translations in Hindi or Marathi as needed (Anderson & Lightfoot, 2018).

The hierarchies of spoken English and fluency in English reflected by the hierarchies of class in Indian society are mirrored in the educational system, with the less socially valued English being spoken in the LFP schools. The *raison d'etre* for the families sending their children to LFP schools, at great cost to themselves, was to learn English as a means toward becoming socially and economically upwardly mobile. Yet, as Bhatia (2018) and Deb (2011) noted, the spoken English the children were acquiring was of a quality unlikely to facilitate admission to the best higher education institutions or access to semi-skilled jobs the families aspired to, such as receptionists, salesclerks or cashiers.

Teachers' Perceptions of the Social Hierarchy of Schools

Universal access to education has created a new hierarchy of access structured on income and social status that is maintained by tuition fees (Chattopadhay & Roy, 2017; LaDousa, 2017). Knowledge of English has become another stratifying factor (Kumar, 2017; Mohan, 2017; Motha, 2014), increasing the social distance between LFP teachers and their students (Arvind, 2015; Sriprakash, 2011) and exacerbating tensions in the status differential. This section examines how class and social hierarchies affects LFP teachers' perceptions of their students and their families.

LFP Teachers' Perceptions of Students and Families

The LFP teachers noted their students came "from slum areas," which they implied affected the quality of the child's education adversely. When asked what they meant by slum areas, the teachers mentioned the neighborhood the schools were in, which included the predominantly Muslim locality and the *basti* with migrant labor. One teacher said in an interview,

> They are all from lower caste. They are all from slum areas. Some are migrant labor, coming from rural areas. They also are living in slum areas.

Although some of the teachers were Muslim themselves, they lived in surrounding neighborhoods, which were more middle-class. The teachers did not send their own children to the same school they taught in (Sriprakash, 2011). Instead, as one noted in an interview, their children went to "very good" English-medium schools that "looked at children's home addresses before deciding to call them in for the admission tests":

> He goes to St. C. School. It's a very good school. You don't get admission. Only my teacher reference is helping him get admission. They don't take children from this neighborhood, only from my neighborhood. In (this region of the city), my son's school is number two. See, it is convent, church school, no?

The first-grade teachers commented that students who had not attended preschool were more likely to fall behind. As one said in an interview, "those are not going to junior-senior (kindergarten), they are not grasping properly," referring to the students' difficulties with comprehension. They felt these children came to school unprepared for its academic rigors, but the parents, being illiterate, did not understand the importance of preschool for their children. Teachers at all grade levels also believed parents needed to be educated about school. One third-grade teacher said in an interview:

> Parents are not understanding. They are coming from slum areas. Many are Muslim. They are illiterate. So, we have to explain, we will teach discipline to your child. Please come to school on time. Please finish your homework. So the parents also learn.

The need to educate the parents as well permeated all aspects of instruction. In an observation, an environmental studies lesson on water storage, for instance, became a didactic opportunity to teach ways to save water at home. The teacher ended the lesson by saying: "Tell your parents, don't throw the drinking water. It hasn't turned bad, we can use it to water the plants." In an interview, another teacher said about the first day of school:

> We had a meeting with the parents and explained to them we will be teaching through games, so we will need some games materials. Because the parents complain that we are always asking them to spend more money, we explained that they don't need to spend any money for these materials. Like these letter cards, they can just make from the tailor's (card)board. We asked them to make four sets of numbers and four sets of each letter. We told them we will ask them to make from low cost materials.

During an observation in a fourth-grade class, a teacher asked two boys, both Muslim, seated at the back of the class, whom she referred to as "superheroes," to come to the front and read aloud a passage from their textbook. When one of them stumbled, she admonished him, "See! You're talking continuously at the back." At this, a student in front called out, "What superhero? He can't read!" In my post-observation interview, I asked this teacher, a Muslim herself, to tell me more about her superhero students. She explained:

> These are my superhero duffers. I have to be very firm with them. They are sitting at the back and they are not concentrating. Instead, they are talking continuously. I have to shout at them. They are coming from the slum areas, and their parents are not educated. They are understanding only the *danda* (stick). They are not understanding about doing homework, bringing the books to school.

The teachers' perspectives on their students' socioeconomic status are rooted in attitudes reflected in society (Jhingran, 2017; Sharma, 2019; Velaskar, 2015). As Jhingran (2017) notes:

> In the Indian situation, bias, and discrimination in the classroom against students on account of caste, religion, gender, language and other identities is common. Such discrimination is not only responsible for lowered self-esteem and alienation, but also constrains the learning process of the children. (p. 87)

Pritchett (2013) described that, when shown how poorly his students were doing, a principal of a village school in eastern Uttar Pradesh reacted with, "We cannot be expected to teach your children. They come from your homes to school stupid and you cannot expect that they will come home from school anything other than stupid." Pritchett added that "the principal had no concern for what his students or their parents thought. He had all the power in the relationship, he knew it, and he was not shy about displaying it" (p. 4). Similarly, Bansal's study (2018) of general education teachers' perspectives on inclusive education noted teachers believed children with special needs tended to come from families from low socioeconomic backgrounds who "lacked the motivation to educate them"; the teachers described the fathers as "alcoholic"; and the parents as being "hardly interested about the performance of their children" (p. 120).

Even food choices have become part of the political discourse on caste and religion "in the context of Hinduism, where vegetarianism is a marker of upper caste identity" (Sathyamala, 2019, p. 878). In one school, where the students were predominantly Muslim and the teachers all Hindu, at break and lunchtimes, I observed the teachers as they went around checking that their students had brought what they deemed as appropriate food items. One teacher berated a student for bringing a sandwich for lunch, since this is traditionally considered a Western fast-food snack item (Aloia et al., 2013). The teacher would ask the class, "*Chapati bhaji* (Indian flatbread and vegetable) is there in your tiffin, no?" and the class would reply in chorus: "Yes, teacher!" One teacher admonished a student, "Mutton *ko laana nahi tha* (you shouldn't have brought any meat). Mummy *ko bolo* (tell your mummy)." Gurumurthy (2021) described similar "instances of Brahminical disciplining," such as a student's "experience of being bullied for bringing eggs in his dabba (tiffin box), being publicly chastised and made to sit on the floor to eat his lunch," and noted:

> The structure of private education, premised on profit, revealed to me, quite unsurprisingly, that all schools (liberal or conservative) maintain and reproduce casteism in similar ways. The deeply pervasive nature of Brahminical ideology, coupled with the political economy of private schools clearly indicate the urgency required in dismantling such an exclusionary system, in order to protect marginalised children from the clutches of Brahminical oppression.

Bhattacharya (2013) also referred to the "putatively secular" curriculum of LFP schools that contained moral messages about cow worship, a cornerstone of Brahmanical ideology.

Although Indian fast food from street vendors are popular snacks among low-income populations (Aloia et al., 2013), the teachers believed "packet snacks" meant uncaring parents. For instance, in an observation during break time, one teacher asked a student what he had eaten. "Ten rupees," replied the student, speaking softly. Later the teacher explained in an interview, as some parents did not have time to make and send homemade breakfast or tea snacks, such as *pohe* or *upma*, with their child, they gave 10 rupees a month to an authorized street vendor to provide these "packet snacks" to the children. As Sathyamala (2019) noted, while vegetarianism in the west is perceived to be both ethical and environmentally friendly, in the "unapologetically" theocratic Indian state of Hindu fundamentalism, eating "transgressive foods" like flesh foods including beef, has emerged as a "political act of subversion" and "a serious challenge to upper caste hegemony" (p. 878).

Teachers' Interactions with Low-Income Parents

The "social distance" (Arvind, 2015) was also manifested in the teachers' interactions with the parents of students who were struggling academically. Like the low-income parents in Ganapathy-Coleman's study (2014), who invoked the cultural concept of the teacher as *guru* as a way to acknowledge the power differential, the parents in this study were submissive, even when the teachers were rude to them. For instance, during an observation of a Saturday remedial lesson, a teacher walked up to the mother of a fourth grader and shouted in Hindi, "He hasn't shown any progress since the beginning of school! Tell me what I should do! He's very slow in writing. Now see how long it is taking him to write! You have to speak to him! Why are you not taking responsibility and asking me, 'what is he learning? How can I help?' You aren't showing any interest in his studies! I am trying to help him by giving him the notes, but I can't give him special attention! I have to help all the children! Why don't you talk to him?".

The mother, a migrant from Bihar, listened quietly to the much younger woman yell at her and then spoke in a reasonable tone of voice. "We have talked to him. I understand what you are saying. What can we do?" Somewhat placated by the mother's submissiveness, the teacher responded less stridently,

> See, if he is not learning, then the blame will fall on me. It is my responsibility that the students learn. So they (the management) will say, 'why are your students not passing?' I want all my students to pass and do well. Here, this is the portion for next week. I will spend two periods on oral review. Two periods is enough. Then for two days, we will have the unit test. If he fails, he can't come to school.

In another instance of an observed remedial lesson, the teacher berated the grandfather, a distinctly older Muslim man, of a third grader. The teacher said, "He's using the same book for everything and he's not making any sense when he writes the notes! Here, I've given him the teacher's notes to write from, but his notes are rubbish!" Once again, the guardian spoke in a soft, appeasing tone of voice, reassuring the teacher that he would make sure the parents bought more notebooks, so the boy would have separate notebooks for each subject.

In only one instance did I observe a brief expression of resistance from a low-income parent. The interaction occurred between two teachers and the father, whose son had missed 6 months of third grade the previous academic year because he had been in an accident and was going to be sent back to the third grade. Speaking softly and calmly, the father requested a refund of his son's tuition for the 6 months the boy had been absent. However, the teachers were firm there would be no refund. When the father pointed out there was a medical reason for his son missing school, one teacher responded, "But it was your responsibility to work with him (while he missed school). You didn't work with him, so he can't go to the fourth standard now." She suggested he appeal to the principal for the tuition refund. At this point, the other teacher jumped in, "If he fails three times, he'll be thrown out of school. Why don't you understand what we're saying?"

The father turned around to her and said, without changing his own tone, "I'm not crazy (*main pagal thoda hoon*) that I don't understand what you're saying. But what I'm trying to say is that he has missed a portion of school and it would be helpful to me if I could get the tuition refunded."

The first teacher reiterated he should approach the principal since fee refunds was not something they as teachers had any control over, and the parent left.

Member checks revealed differences of opinion about these interactions between the teachers and low-income parents. Upper-income, westernized professionals decried the behavior, although all agreed it was common. However, other professionals felt this was justified. For instance, a university professor said:

> I'm not blaming the teachers. If you shout, they (the parents) understand you are showing concern. Depends on how parents are behaving. Possibly there is no response. The attitude is that if parents are not paying attention, you have to shout and talk loudly. It's the cultural mentality. It is a sanctioned, approved role, that the teacher is the expert. (If) I am the parent, I am responsible to listen to you. Can't say, I am young or old, that is no excuse if the parent is not paying attention. Sometimes the fathers who are dominant in society might question, who are you to talk like this? This is a gender bias. And when a male teacher will shout, they will listen.

In the situations described, despite being younger than the parents and grandparent, which traditionally would have required them to show more respect, teachers exerted their higher status (Ganapathy-Coleman, 2014). As Limaye (2019) noted, "generally in Indian schools, parents are expected to listen to the schools' officials and follow their instructions… The parents do not have a right to challenge the school's decisions" (p. 654). Further, being first-generation English speakers gave the teachers an edge over their students' parents.

Low-Income Parents' Position in the Social Hierarchy

Interviews with parents indicated an acute awareness of this status differential along with a recognition that they had little recourse to make changes within their current societal context. As one mother said:

> What can we do? What can we say? We are helpless (*hum majboor hain*). If we shout back at the teacher, she will say, 'if you don't like the way I teach, take your son out of the school.' I am illiterate. How can I say anything to her?

Disproving the teachers' attitudes toward them as being uncaring, the parents demonstrated deep caring for their children. They were willing to undergo great financial hardship by sending their children to these schools because they saw English-speaking skills as a pathway to their children's social mobility.

They also indicated they were learning how to navigate the system in their favor. For instance, one parent mentioned that because they had not known the importance of preschool or kindergarten, their older son had failed the first-grade entrance exam to the English-medium school. With the second son, they were more prepared and made sure he went to "junior-senior," which, she asserted, had helped him get admission to the English-medium school. Similarly, many parents said that although they could not help their children with their studies, their children were taking private tuition classes and the tuition teacher would support them. As one parent noted:

> I can't help my son, but I know that he is working very hard at his studies. He goes to tuition where the teacher helps him complete his homework, review the lesson, understand what was happening in school, everything. So even though I can't help him, the tuition teacher makes sure that he is learning. If M comes home sometimes without his notes, she makes sure to follow up to get them.

Thus, the tuition teacher became a sort of surrogate parent support for them and the teacher, "the bridge between school and home" as a tuition teacher put it. All the parents expressed faith that the difficulties their child was experiencing in school, which the class teacher had brought to their notice, would ease with the tuition teacher's support and as the child progressed through the grades and became more comfortable with English. They described the difficulties in terms of being "slow in writing" or "*thoda* (a little) weak." As one father stated:

> I think he's doing fine. His handwriting is a little sloppy (*vesey-tesey*) but he seems to understand what is happening. I didn't get an education so I want him to be educated and be able to do well. Nowadays he will need English, to speak it, to write it, to read it.

Conclusion

The subtle and overt hierarchies that abound within the larger social context and permeate almost every aspect of Indian life are perpetuated by attitudes toward a multitude of variables, such as caste, gender and religious difference (Jhingran, 2017; Sharma, 2019; Velaskar, 2015). As Yengde (2019) asserted:

> While the economic system of feudalism has been gradually replaced with a mixed modern-day model of capitalism and state socialism, the social aspect of feudalism is as entrenched as ever. The inherent form of capitalism reproduced with it the age-old structure of oppression. (p. 28)

As Avramidis and Norwich (2002) noted, teacher attitudes are context dependent, "arising out of interactions with others… and responsive to factors within a particular sociocultural environment" (p. 144). Thus, since the schools, the teachers' "habitus" (Bourdieu & Passeron, 1990), are imbued with social hierarchies and constructs of caste/class, the teachers' attitudes reflect both new and old structures of oppression.

The tiered system of the schools was maintained by tuition costs, ensuring which social class of students could attend. Upper-tier schools had resources that were like schools in the US, while the LFP schools had minimal resources. Tiers of fluency in English also affected social status, maintaining the social class of the upper-tier teachers, which was on par with their students. On the other hand, the LFP teachers sought to distance themselves socially from their students, whom they saw as coming from slum areas, with illiterate and uncaring parents. Although poverty plays a major role in denying them an education, the study found that low-income parents were less likely to question an ineffective education system (Ganapathy-Coleman, 2014) and accepted its indignities and inadequacies of the educational system as they sought to escape their poverty.

The benevolent intention behind universal access to education has been subverted by neoliberal reforms that focus on productivity and cutting costs, transforming the teaching profession, especially for women (Belliappa, 2014; Manjrekar, 2013). Colliding with age-old societal hierarchical traditions, as more poor children enter schools, the social distance between teachers and students has increased, shaping teachers' perceptions of their students and their families in LFP schools (Arvind, 2015).

The recent and highly desirable social asset of the teachers' knowing English further exacerbated this social distance. They lack agentic control to make change, but, more significantly, they see no need to make change. This is the tension of their lived reality, where they must negotiate their work status within the confines of their personal lives and fulfill their domestic responsibilities, and still be able to expect, even demand, some measure of respect as professionals from the parents of their students from the "slum areas."

Their attitudes, however, indicate a significant fault line in the envisioning of quality education for all within the current context of neoliberal education reforms. The difference in the quality of education that emerges from private schools for affluent students and for low-income students points incontrovertibly to an inherent contradiction in the expectation that the private sector, bound as it is by economic constraints of profit, competition and productivity, no matter how philanthropic the enterprise, is the solution to the government's failure (Chattopadhay & Roy, 2017; Sarangapani & Winch, 2010). The belief everyone will benefit equally merely by making education available to all overlooks the ineluctable fact that school systems reflect the larger global and national influences (Stromquist & Monkman, 2015) and participate in the cultural reproduction of existing social and economic hierarchies (Bourdieu & Passeron, 1990; Collin & Apple, 2010).

The next chapter examines the complexities of English as a medium of instruction for first-generation English learners. It presents the challenges teachers and students in LFP schools experience when students are both learning in English and learning English simultaneously. Lacking the pedagogical skills for systematic bilingual and literacy instruction, teachers resort to teaching through rote memorization and begin to conflate difficulty with learning language with difficulty with learning. In comparison, students in high-tier schools who are fluent in English participate in enriched educational activities and are further privileged.

Notes

1. As Bhattacharya (2013) notes, this statistic is uncertain and other sources cite different figures. The point is that, whatever the actual figures, the numbers of English-speakers continue to be in a minority.

2. Canagarajah (2002) asserts the need to recognize these varieties of English that have emerged from former colonies as World Englishes in their own right as a way of resisting the linguistic imperialism and hegemony of "inner circle" speakers of English.

References

Aloia, C. R., Gasevic, D., Yusuf, S., Teo, K., Chockalingam, A., Patro, B. K., Kumar, R., & Lear, S. A. (2013). Differences in perceptions and fast food eating behaviours between Indians living in high- and low-income neighbourhoods of Chandigarh India. *Nutrition Journal, 12*, 4. https://doi.org/10.1186/1475-2891-12-4

Ajmera, M. (2016). *Invisible children: Reimagining international development at the grassroots*. Palgrave Macmillan.

Anderson, J., & Lightfoot, A. (2018). Translingual practices in English classrooms in India: Current perceptions and future possibilities. *International Journal of Bilingual Education and Bilingualism*. https://doi.org/10.1080/13670050.2018.1548558

Annamalai E. (2004). Medium of power: The question of English in education in India. In J. W. Tollefson & A. B. M. Tsui (Eds.), *Medium of Instruction policies: Which Agenda? whose Agenda?* (pp. 177–194). Routledge.

Arvind, G. R. (2015). Institutional context, classroom discourse, and children's thinking: Pedagogy re-examined. In M. Thapan (Ed.), *Education and society: Themes, perspectives, practices* (pp. 283–308). Oxford University Press.

ASER. (2019). *Annual status of education report, 2018*. Pratham, Anthem Press.

Avramidis, E., & Norwich, B. (2002). Teachers' attitudes towards integration/inclusion: A review of the literature. *European Journal of Special Needs Education, 17*(2), 129–147.

Babu, M. J. P. (2018). Facilities provided by the primary education schools—A study in east Godavari district of Andhra Pradesh. *International Journal of Academic Research and Development, 3*(2), 1605–1607.

Bandura, A. (1999). Social cognitive theory: An agentic perspective. *Asian Journal of Social Psychology, 2*(1), 21–41.

Bangay, C., & Latham, M. (2013). Are we asking the right questions? Moving beyond the state vs non-state providers debate: Reflections and a case study from India. *International Journal of Educational Development, 33*, 244–252. https://doi.org/10.1016/j.ijedudev.2012.09.004

Bansal, S. (2018). Understanding Teachers' Perspective of Inclusive Education for Children With Special Needs (CWSN). *Educational Quest: An International Journal of Education and Applied Social Science, 9*(1), 115–123.

Belliappa, J. L. (2014, October 15). *Gender inequities in school teaching: Reasons and repercussions*. Learning Curve. http://www.teachersofindia.org/en/article/gender-inequities-school-teaching-reasons-and-repurcussions

Bhatia, S. (2018). *Decolonizing Psychology: Globalization, Social Justice, and Indian Youth Identities*. Oxford University Press.

Bhattacharya, U. (2013) Mediating inequalities: Exploring English-medium instruction in a suburban Indian village school. *Current Issues in Language Planning, 14*(1), 164–184. https://doi.org/10.1080/14664208.2013.791236

Boo, K. (2012). *Behind the beautiful forevers: Life, death and hope in a Mumbai undercity*. Random House.

Boruah, P. (2017). Learning English in a low-cost semi-urban English-medium school in India: Challenges, interaction patterns and domains of use. In H. Coleman (Ed.), *Multilingualisms and Development* (pp. 289–306). British Council.

Bourdieu, P., & Passeron, J. C. (1990). *Reproduction in education, society and culture*. Sage.

Center for Equity Studies (CES). (2014). *India exclusion report, 2013–14*. Books for Change.

Chabbott, C. (2015). *Institutionalizing Health and Education For All: Global goals, innovations and scaling up*. Teachers College Press.

Chattopadhay, T., & Roy, M. (2017). *Low-Fee Private Schools in India: The Emerging Fault Lines* (Working Paper 233). National Center for the Study of Privatization in Education. Teachers College, Columbia University. http://ncspe.tc.columbia.edu/center-news/low-fee-private-schools-in-india/

Collin, R., & Apple, M. (2010). Globalization, schooling, and literacies in an unsettled age. In S. Tozer, B. P. Gallegos, A. Henry, M. Bushnell Greiner, & P. Groves Price (Eds.), *Handbook of research in the social foundations of education* (pp. 296–307). Routledge.

Datt, G., & Wang, L. C. (2020). Schooling and learning deficits: A simple unified measurement framework. *Comparative Education Review, 64*(3). http://doi.org/0010-4086/2020/6403-0008

Day, L. C., McLoughlin, C., Aslam, M., Engel, J., Wales, J., Rawal, S., Batley, G., Kingdom, S., Nicolai, S., & Rose, P. (2014). *The role and impact of private school in developing countries: A rigorous review of the evidence*. Department for International Development.

Deb, S. (2011). *The beautiful and the damned: A portrait of the new India*. Faber.

Ganapathy-Coleman, H. (2014). 'Teachers are always good. Children have flaws': Memories of school and learning in the narratives of low-income parents in India. *Psychology and Developing Societies, 26*(1), 29–58.

Guha, R. (2020, June 9). *Traits of Modi that have cost India dearly*. NDTV. https://www.ndtv.com/opinion/3-traits-of-modi-that-have-cost-india-dearly-by-ramachandra-guha-2242669

Gurumurthy, A. A. (2021, May 28). What I learned about caste while discussing the PSBB incident with family, friends. *The Wire*. https://livewire.thewire.in/campus/psbb-incident-casteism-oppression/

Heyneman, S. P., & Stern, J. M. (2014). Low cost private schools for the poor: What public policy is appropriate? *International Journal of Educational Development, 35*, 3–15.

Hindi mother tongue of 44% of Indians, Marathi 3rd most spoken. (2018, June 27). *Times of India*, p. 11.

Jhingran, D. (2017). Education of equitable quality: All children reading. In M. J. Tiwary, S. Kumar, & A. K. Mitra (Eds.), *Dynamics of inclusive classrooms: Social diversity, inequality and school education in India* (pp. 81–104). Orient BlackSwan.

Juneja, N. (2017). Education in urban areas. In K. Kumar (Ed.), *Routledge handbook of education in India, debates, practices, and policies* (pp. 26–42). Routledge. https://www.routledgehandbooks.com/doi/10.4324/9781315107929.ch2

Khubchandani, L. M. (1975). Dilemmas of language transition: Challenges to language planning in India. *ERIC Document Reproduction Service No. ED, 116*, 474.

Kumar, N. (2017). The technology and politics of teaching English. In M. J. Tiwary, S. Kumar, & A. K. Mitra (Eds.), *Dynamics of inclusive classrooms: Social diversity, inequality and school education in India* (pp. 183–202). Orient BlackSwan.

LaDousa, C. (2017). Mind the (language-medium) gap. In K. Kumar (Ed.), *Routledge handbook of education in India, debates, practices, and policies* (pp. 81–96). Routledge. https://www.routledgehandbooks.com/doi/10.4324/9781315107929.ch5

Limaye, S. (2019). Unheard voices: Schooling experiences of parents having children with disabilities in India. In M. J. Schuelka, C. J. Johnstone, G. Thomas, & A. J. Artiles (Eds.), *The SAGE Handbook of diversity and inclusion in education* (pp. 557–570). Sage.

Manjrekar, N. (2013). Women school teachers in new times: Some preliminary reflections. *Indian Journal of Gender Studies, 20*(2), 335–356.

Menon, S. S. (2019). *Stakeholder perspectives on teacher attrition in private early childhood schools in India* (Unpublished dissertation). Walden University.

Ministry of Human Resource Development. (2019). *Draft National Education Policy 2019*. Government of India.

Ministry of Law and Justice. (2009). *The Right of Children to Free and Compulsory Education Act*. Legislative Department. https://www.mhrd.gov.in/sites/upload_files/mhrd/files/upload_document/rte.pdf

Mohan, P. (2017). The road to English: The slow migration of the EWS child to elite India. In M. J. Tiwary, S. Kumar, & A. K. Mitra (Eds.), *Dynamics of inclusive classrooms: Social diversity, inequality and school education in India* (pp. 203–215). Orient BlackSwan.

Motha, S. (2014). *Race, empire, and English language teaching: Creating responsible and ethical anti-racist practice*. Teachers College Press.

Mukhopadhyay, L. (2020). Translanguaging in Primary level ESL classroom in India: An exploratory study. *International Journal of English Language Teaching, 7*, 2. https://doi.org/10.5430/ijelt.v7n2p1.

NCERT. (2005). *National Curriculum Framework*. Author.

Pandey, G. (2011, February 15). *An 'English goddess' for India's down-trodden*. BBC News. https://www.bbc.com/news/world-south-asia-12355740

Pritchett, L. (2013). *The rebirth of education: Schooling ain't learning*. Brookings Institution Press.

Sarangapani, P. M., & Winch, C. (2010). Tooley, Dixon and Gomathi on private education in Hyderabad: A reply. *Oxford Review of Education, 36*(4), 499–515. https://doi.org/10.1080/03054985.2010.495465

Sathyamala, C. (2019). Meat-eating in India: Whose food, whose politics, and whose rights? *Policy Futures in Education, 17*(7), 878–891. https://doi.org/10.1177/1478210318780553

Sharma, R. (2019, February 3). The class divisions that won't fade. *The New York Times*, p. 6.

Shenoy, S., & Wagner, R. K. (2019). Language and literacy practices that influence bilingual and bi-literate acquisition in L1 Kannada and L2 English in Bangalore, India. In R. M. Joshi & C. McBride (Eds.), *Handbook of literacy in Akshara orthography* (pp. 373–388). Springer. https://doi.org/10.1007/978-3-030-05977-4_19

Singal, N., Johansson, S. T., & Lynch, P. (2019). Education of children with disabilities: Changing landscape of new opportunities and challenges. In N. Singal, S. T. Johansson, & P. Lynch (Eds.), *Education and disability in the global South: New perspectives from Africa and Asia* (pp. 1–18). Bloomsbury Academic.

Sriprakash, A. (2011). Being a teacher in contexts of change: Education reform and the repositioning of teachers' work in India. *Contemporary Education Dialogue, 8*(1), 5–31.

Stromquist, N. P., & Monkman, K. (2015). Defining globalization and assessing its implications for knowledge and education, revisited. In N. P. Stromquist & K. Monkman (Eds.), *Globalization and education: Integration and contestation across cultures* (2nd ed., pp. 1–20). Rowman & Littlefield.

Tooley, J. (2009). *The beautiful tree: A personal journey into how the world's poorest people are educating themselves*. Penguin.

Upadhyay, S. B. (2012). Dalits and the modern education in colonial India. In C. Sleeter, S. B. Upadhyay, A. K. Mishra, & S. Kumar (Eds.), *School education, pluralism and marginality: Comparative perspectives* (pp. 152–180). Orient Blackswan.

Velaskar, P. (2015). Educational stratification, dominant ideology, and the reproduction of disadvantage in India. In. M. Thapan (Ed.), *Education and society: Themes, perspectives and practices* (pp. 428–448). Oxford University Press.

Vulli, D. (2014). English and medium of instruction: Dalit discourse in Indian education. *Research Journal of Educational Sciences, 2*(2), 1–6.

Waldrop, A. (2015). The meaning of the old school tie: Private schools, admission procedures, and class segmentation. In M. Thapan (Ed.), *Education and society: Themes, perspectives and practices* (pp. 405–427). Oxford University Press.

Wagner, D. A., Wolf, S., & Boruch, R. (2018). *Learning at the bottom of the pyramid: Science, measurement and policy in low-income countries*. UNESCO.

Yengde, S. (2019). *Caste matters*. Penguin Viking

CHAPTER 5

Language and Learning in the Classroom

The previous chapter examined how the teaching profession, especially for women, has changed as the neoliberal focus on productivity and economy collided with age-old societal hierarchies. The social distance between parents and teachers in the LFP schools is aggravated by the language factor—that the teachers know English, a highly desirable social asset, and the families do not. However, English being the medium of instruction (MOI) only serves to add an additional layer of complexity to students' learning problems in LFP schools and often results in the conflation of cognitive deficit with language difference. This chapter focuses on these two themes: How the complication of English as MOI results in and reinforces rote learning and teaching to the test, and how students' struggles with acquiring English are perceived to be learning difficulties. It argues that this emphasis on rote memorization and the "marks race" (Majumdar & Mooij, 2012, p. 207) subverts the purpose of education as a development goal of an engaged, discriminating citizenry and generates an illusory meritocracy that further advantages the privileged and disables students already disadvantaged by lack of access to English.

M. Kalyanpur, *Development, Education and Learning Disability in India*, Palgrave Studies in Disability and International Development,
https://doi.org/10.1007/978-3-030-83989-5_5

The Complication of English as a Medium of Instruction

The immersion model of education, where all instructional content is taught in a single language (except for languages that are taught as a subject), has been the primary structure of educational systems (García, 2019; Shenoy & Pearson, 2018). While this model has been successful for students who speak the same language at home as is being taught in the school, it has proven to be less successful with students who speak a different home language (L1) from the MOI in school (L2). Children use language for basic interpersonal communication skills, or BICS, and through school acquired cognitive academic language proficiency, or CALP (Cummins, 1979, 1981; Skutnabb-Kangas, 2017). However, students who have not yet mastered oral and written fluency in their L1 before being introduced to L2 often struggle to acquire both and become proficient in neither (Cummins, 1979). Another scenario is where students may maintain their L1, and successfully acquire L2, but over time use their L1 only for BICS and lose their CALP in a process of subtractive bilingualism (Skutnabb-Kangas, 2017). In other words, the immersion model of education that predominates educational systems disadvantages bi- or multilingual speakers, who constitute the majority of children in India (Shenoy & Pearson, 2018). The percentage of native English speakers in India is very small, and most children come to school speaking a language other than English; low-income children, in particular, have no or very limited exposure to it (Boruah, 2017). Although instructional practices are similar for both groups, a given practice might prove less effective for English language learners because they face the double challenge of learning academic content and the language of instruction simultaneously (Goldenberg, 2008, as cited in Shenoy, in press).

Further, even traditional bilingual programs do not acknowledge the "fluid language practices [of multilingual speakers who might] simply deploy all the features of [their] repertoire... suppressing some linguistic features and selecting others" (García, 2019, p. 164). Code-switching, where students are encouraged to code-switch by applying the linguistic norms of the language appropriate to the listener's context, has been found to be a successful basic translanguaging practice in bilingual education (García et al., 2017). For instance, Mukhopadhyay's (2020) study of one teacher's formal application of translanguaging in an Indian primary government-run school, involving the systematic use of Hindi and Telugu to introduce academic language in English, showed students were more likely to engage in academic discourse and develop academic language proficiency than if taught in the strict monolingual mode.

However, the LFP teachers in this study tended to use "teaching-in-translation" (Bhattacharya, 2013, p. 173), which is not always easy or systematic (Anderson & Lightfoot, 2018; Mohan, 2017). Similarly, Bhattacharya's (2013) study found "texts were translated and paraphrased into Hindi without pointing out which syntactic and lexical items were being introduced or excluded in the translation process [which] affected students' ability to identify the meaning of individual words" (p. 173). As this section shows, merely providing translations for the content from English to students' regional languages, as the LFP teachers do, neither provides a deeper, nuanced understanding of the languages being used (Bhattacharya, 2013), nor accounts for the complexity of the variety of scripts involved within the Indian context (Mukhopadhyay, 2020).

Teaching English and Teaching in English in the LFP Schools

English as MOI became an overwhelming barrier in the LFP schools in the study because students had neither BICS nor CALP skills in English. They came to school to learn English, even as they were being asked to learn in English (Bhattacharya, 2013; Boruah, 2017). As in Boruah's (2017) study, spontaneous use of English among the students was limited to utilitarian statements like "Please can I enter, ma'am?" or "Please can I go to toilet, ma'am?" Although some students could answer teachers' questions on the content in English, in most cases, the teacher had to remind the students to answer in English, at which point the student would become silent. Bhattacharya's (2013) and Mohan's (2017) studies separately noted a similar silencing among students struggling to express themselves in English. Students' unfamiliarity with both basic and academic English vocabulary affected their conceptual comprehension, leaving "a whole universe of meaning… hidden from them" (Mohan, 2017, p. 21).

The Need to Translate Basic and Academic Vocabulary Words

Although LFP teachers in the study were aware their students were learning English as a new language, they were always conscious of the schools' English only policy. In an interview, one teacher stated:

> See, they are coming to this school to learn English. So we have to teach them. If they are having problems, they should go to Marathi-medium [school].

Yet, in an effort to help students, they used a mixture of languages to explain basic words, consistently engaging in this "guilty multilingualism" (Coleman, 2017, p. 31) as they provided "teaching-in-translation" (Sriprakash, 2011) for their English language learners. My fieldnotes of an observation of a lesson on animal shelters show the teacher using English, Hindi and Marathi:

> "Where does ant live?" asked K in English, and then added '*cheentee*' and '*mungi*', the words for ant in Hindi and in Marathi respectively.

The inclusion of academic words like "depleted," "potable," and "contaminated" water, and "economically," as in water must be used economically, in a fourth-grade textbook on environmental sciences used in an LFP school in this study points to Bhattacharya's (2013) observation that "the disconnect between the content of various textbooks and students' everyday lives posed a barrier in the acquisition of language" (p. 175). While students were very familiar with the *concepts* of potable and contaminated water and the need to be economical because they lived near open drains with limited access to clean water, they were unfamiliar with the academic English words for them. Similarly, in her study, Bhattacharya (2013), who found vocabulary words in textbooks such as "library," "computer," "swimming pool," "peon," and "maid," noted LFP students had not encountered computers and swimming pools. She asserted they were familiar with peons and maids; however, not as people who rang schoolbells and looked after children at schools, but in the livelihoods of their own parents. As Mohan (2017) stated,

> The abstruseness of the prose in English comprehension passages, with obscure allusions, and the needless complexity of the explanations in science texts ensure that teachers and children will end up memorising set answers and glossing over chunks of mysterious stuff, instead of engaging with the language. (p. 22)

Losing Conceptual Understanding Amidst the Translation

Students' unfamiliarity with English became a major stumbling block to conceptual understanding, even when teachers used more innovative approaches to teach reading, such as phonics, instead of the traditional Alphabet-Spelling approach, in which children are taught the letter names and then spell out words (Gupta, 2014). This example of a phonics lesson I observed, where the teacher was introducing students to letter sounds, demonstrates the levels of complexity emerging from this unfamiliarity with basic and academic vocabulary. As my fieldnotes show:

> Speaking entirely in Hindi except for the word "blending," the teacher asked, "What does 'blending' mean?" She waited and, when no one answered, gave the answer herself in English, "Blending is joining," adding the word, *jodna* in Hindi. Pointing to word pictures, the teacher then reviewed a phonics song which went through the alphabet in order and introduced them to letter sounds, associating the sounds with words like "igloo" and "gorilla." The class sang along with enthusiasm.

Later when I asked them to name the picture words in the alphabet song, several students were unable to do so and some gave the Hindi word for the item. For instance, one student said, "M for *bandar*," which is Hindi for monkey. When I asked if words more familiar to the students, such as "India" for the letter sound "eye," might help them remember the association more easily, the teacher pointed out, rightfully so, she was not the one setting the curriculum, stating,

> Management is giving us training how to use, how to teach. We are just following. They are giving us the materials, so I am using them.

Here, students' unfamiliarity with the words to illustrate the phonic sounds complicated their comprehension of the *concept* of letter sounds which, in turn, compromised their ability to read using a phonics approach. In this situation, neither the academic term "blending" nor the basic word "joining" to explain the academic term was familiar to students and needed translation. Further, although the teacher may have recognized these difficulties for her students, she had no strategies to nuance her instruction. In my interview with Dr. Nandini Singh, then at the National Brain Research Center, she noted in her observations of classrooms:

> I saw that the teachers knew nothing about how to teach reading. This unsystematic instruction meant that kids were not learning orthographic or phonetic rules that would help them to learn how to read.

Similarly, in her study of phonics implementation in a rural school, Gupta (2014) found, although children were given the mother tongue equivalent of the English words beneath the pictures, the focus was on chanting the sounds, so students did not become familiar with the words. Noting teachers struggled with teaching letters in English that had two or more sounds, Gupta (2014) asserted teachers "domesticated the pedagogy to suit local constraints" (p. 3911), applying the schema they used to teach Indic languages, which are phonetic. Importing instructional approaches unsystematically and without consideration of familiar systems of teaching techniques, Gupta concluded, subverts outcomes for both students and teachers.

Differences in Scripts as an Additional Barrier

Most students in this study were not necessarily biliterate, which, Shenoy and Pearson (2018) assert, is an additional barrier. This was complicated by the variety of scripts involved in the different languages they spoke (i.e., Roman for English, Devanagari for Hindi and Marathi and Arabic for Urdu). Given the meaning of a word in a language they understood, but lacking access to or fluency in its script, many could not write down the meanings of English words in their preferred language in their textbooks. For instance, during an observation of a lesson in the second-grade, the teacher explained the meaning of fort as *kila* in Hindi, which they were familiar with. However, because the second graders could not write Hindi, they had no other recourse but to commit the meaning to memory. In the upper grades, only a few students were fluent enough in Hindi to write the translations in their textbook before the teacher moved on, and most students resorted to memorization. In some situations, the older students transliterated, writing the Hindi word in the Roman script, which is becoming an increasingly popular way to access Hindi (Almeida, 2020).

Missing the Conceptual Understanding in Students' Responses

While teachers consistently applied teaching-in-translation in their own instruction, they seldom "encouraged the development of [their students'] multilingual competence" (Anderson & Lightfoot, 2018, p. 16). Students' responses in Hindi were not accepted, as were partially correct responses in English, even if they demonstrated comprehension of concepts. In one instance, during an observation of a lesson on English grammar, when asked if he understood the meaning of the sentence,

"He poured the oil from a height," a student answered, *lambai*, the Hindi heteronym for height, as in tallness—most likely the context in which he had heard the word. Similarly, in an observation of an environmental studies lesson, a teacher asked the students what they might use well water for. "Clothing wash," answered one girl in English, offering a grammatically literal translation from Hindi, in which the noun precedes the verb. Although both answers suggested conceptual understanding, teachers made "no attempt to clarify the propositional mis-construction" (Arvind, 2015, p. 273), or the logic underlying students' thinking.

Anderson and Lightfoot (2018) suggested teachers' responses may be related to the ostensible English only policy, which creates "the pressure to [pretend to] teach only in English for some, disdain for hybrid languaging practices for others, and conservative curricula and assessment criteria" (p. 16). Indeed, when I asked, the teachers clarified that the answers were incorrect because they were not in English or grammatically accurate, insisting, in the words of one: "Now they are in third standard, they must give answers in English. This is English-medium school." Similarly, a representative of a non-governmental organization (NGO) that supports and promotes English-medium instruction in LFP schools explained in an interview:

> Essentially, even if you say the problem is English, what other language is there? There's probably ten kids in each class that speak the same language. The teachers don't necessarily speak it either. So the common language, in the end, is English. All the parents want is for their child to learn English. And even the resources for teacher education are universally available in English.

Although their point about multilingual contexts is well-taken, the fact is outcomes for students in English-medium LFP schools are very poor (Anderson & Lightfoot, 2018; Bhattacharya, 2013; Boruah, 2017; EPW Engage, 2021): From the 190 English-medium LFP schools this NGO supported at the time of my study, only one student had received admission into a top-tier secondary school and college.[1] As Groff (2017) noted, "children who do not fully understand the language of instruction may be learning a dominant language by submersion but miss out on content learning" (p. 136).

Research evidence emphasizes the importance of mother tongue instruction in the context of bilingual/multilingual education as most beneficial to learners from minority language groups (Groff, 2017); as UNESCO (2016) asserted, this approach "facilitates learning both of academic subjects as well as the second language (L2). At school, children can concentrate on the subject matter instead of struggling to understand the language of instruction as well" (p. 2). Instead, because of English only policies, the quality of education was severely compromised for students in these LFP schools and the intent of EFA subverted.

Language and Learning in Upper-Tier Schools

The starkness of the quality of education the LFP students received because of lack of familiarity with English became even more apparent when compared to the educational quality for upper-tier school students. In these upper-tier schools, too, all instruction was in English. However, in addition to the classroom resources described in Chapter 4, the most significant advantage the teachers had was the students' familiarity with the language in which they were being taught. Not having to constantly translate into another language—and therefore not having to think simultaneously in at least two languages—saved these teachers enormous effort and time. By being able to assume students' facility with basic vocabulary in English, they were able to focus on academic vocabulary and expand students' learning far beyond the textbook. For instance, in an observation of a lesson on grammar, in the same amount of time as a lesson in the LFP schools, third-grade students in an upper-tier school correctly paired words to match adjective to noun, (e.g., adorable baby, sleeveless dress), named the academic vocabulary (e.g., adjectives, nouns), and made sentences using these pairs of words in a round-robin (i.e., students taking turns to build on the previous sentence to make a story). When asked the purpose of adjectives, one student replied in fluent English, "To make the sentence more interesting," and another student added, "It describes something so the reader can understand it better." The teacher showed students four adjectives (scary, tasty, beautiful and big) on the Smartboard, and then placed the students in small groups to match their meaning with additional words from the teacher (which included vocabulary like towering, glamorous, colossal, spine-chilling, appetizing, gigantic, petrified). Finally, the teacher gave students the option of creating a song, drawing a descriptive scene, or writing a role play with the instruction that they "use as many adjectives as you can to show your understanding." As these fieldnotes illustrate, although the teacher used the Smartboard and incorporated interactive activities to teach this lesson, the students' ability to respond meaningfully, immediately comprehend and apply their learning was also made possible by the fact that they were very comfortable with the language in which they were learning.

In this context, the thorniness of the language issue becomes apparent (LaDousa, 2017; Bhattacharya, 2013; Chattopadhay & Roy, 2017; Mukhopadhyay, 2020; Shenoy & Pearson, 2018). Despite low-income parents' belief that going to an English-medium LFP school afforded their children an additional fillip in their education, the MOI ended up becoming the most significant barrier to the quality of their education. It illustrates the challenge of teaching monolingually in a multilingual context, amid constraints of teachers poorly qualified to teach reading, limited classroom resources, and large class sizes, providing teachers the rationale to move towards rote teaching as the most efficient instructional method. By contrast, the upper-tier schools gave students already advantaged by their facility with English further access to an enriched curriculum. The upper-tier students had opportunities for critical thinking through written and oral expression that would facilitate their admission to higher education nationally and internationally, perpetuating the cultural reproduction of school systems and the hegemony of the elite (Collin & Apple, 2010; García, 2019).

Using Rote Memorization in Child-Centered Teaching

In keeping with LFP schools' preference to hire less prepared teachers, and then teach them on the job (Tooley, 2009), the LFP schools in this study had arranged for teachers to receive in-service professional development. Most of this training focused on what the teachers called "child-centered teaching," which, Sriprakash (2011) noted, has become part of quality pedagogical improvement initiatives that have "broadly sought to shift the teacher's role from that of an instructor to that of a 'facilitator' of learning" (p. 6). The teachers in the LFP schools did make efforts to use child-centered teaching. However, the complexities of language combined with the pressures of finishing the curriculum, large class sizes and few material resources often compelled teachers to fall back on traditional teacher-led approaches (Gupta, 2014), such as rote learning and copying from the board, as being the most expeditious means of teaching to the test (Nawani, 2015).

Child-Centered Teaching

In one school, the teachers mentioned for the past 2 years before school started they had attended training workshops organized by management, where they learned about "play aways" (Tooley, 2009, p. 14). They spoke about using games to teach, what they called "hands activities" or "teaching games," and brought out teaching materials and manipulatives they had developed at trainings. Some had enlisted parents to send low-cost materials, like flashcards from cardboard, matches and beads.

Most modifications were simple, requiring no additional resources. One teacher used colored chalk to color-code and chunk information on the board. Another teacher used her fingers to teach about geometric angles, where a V sign was an acute angle, an L sign was a right angle, and a raised pinky and thumb was an obtuse angle. Several teachers introduced some physical movement to transition between class periods, such as sing-alongs with hand gestures, exercising in place, playing "Simon says."

I observed some teachers use the gradual release of responsibility framework (Frey & Fisher, 2009), a 3-step process often called "I do, we do, you do." The teacher first modeled a concept or construct on their own, then guided the students through the construct together, and finally let the students apply their learning independently. Some teachers combined this strategy with using manipulatives. In post-observation interviews, teachers showed a level of sophistication in their ability to reflect on the pedagogical strategies they used, although they did not use the academic language of gradual release or manipulatives. One teacher described in an interview:

> First, I am showing them how to do, then I am having them practice. For maths, I try to use teaching games. Here I am doing the bead game. In the training workshop we learnt about using teaching games. The students only are bringing the beads.

A popular language arts teaching game that I observed was the exercise the teachers called "Who said to whom?" In the game, teachers used the dialog in the text to navigate the students through the plotline of a story. In one observation of an English grammar lesson, the teacher reviewed an Aesop's fable prior to the practice test by having a few students stand up in turn and recount the story through the script written on the board. Later in a post-observation interview, she explained this structure helped students to remember the sequence of the story and improve their English.

All teachers in the study described these strategies as part of child-centered teaching. However, they admitted they could not use teaching games on a regular basis because of the constraints of teaching the curriculum and the large class sizes. Scholars note teachers often experience a tension between applying child-centered teaching and the pressures of being accountable for completing a pre-specified curriculum (Arvind, 2015; Bhatnagar & Das, 2014; Mili, 2018; Sriprakash, 2011). For instance, in Sriprakash's study (2011), teachers found it easier to teach the syllabus content directly, rather than implement the prescribed activities of child-centered teaching and wait for children to learn. The challenges of juggling several languages, as the previous section showed, provide a further rationale for teaching by rote.

Even when teachers allowed students to be creative, the parameters for creativity were limited, and, as Boruah (2017) also noted, students themselves were so conditioned to being told what to write, that they had difficulty being very creative. For instance, in a grammar lesson that I observed, the teacher gave her students sentence starters, like "The flowers are...," then allowed them to come up with their own ideas for completing the sentence. The students appeared to enjoy this. One girl said, "*Mazaa ah raha heh!*" (This is fun!), completing her sentence starter "Delhi is..." with "a big city." Then, the teacher began to solicit ideas. "Delhi is the capital of India," suggested another student, to which the teacher said, "Very good sentence! Clap, everybody!" At this, the little girl who had expressed her enjoyment at coming up with her own ideas erased her original sentence and wrote the one she believed the teacher had selected for them. Thus, Boruah (2017) noted, English-medium instruction becomes "a pseudo-immersion exercise that does not ensure grammatical competence and does not ensure development of originality and creative use of English" (p. 301). Arvind (2015) reflected even seemingly innovative approaches often hide behind traditional techniques like rote learning and do not facilitate conceptual understanding. For instance, both the gradual release of responsibility math and the language arts lessons observed used the conventional didactic drill-and-memorize approach teachers often fell back on.

Emphasis on Rote Memorization

Foundational rote-based learning, such as multiplication tables or phonics toward acquiring fluency in numeracy and literacy, has merit when combined with inquiry-based learning such as constructivism for developing critical thinking skills (Arvind, 2015; Soler, 2016). The problem arises when training in child-centered teaching notwithstanding, rote memorization becomes the primary instructional method used (Arvind, 2015; Sriprakash, 2011). For teachers in the study, however, the many constraints they faced with regard to language challenges, large class sizes, specified curriculum and minimal classroom resources, became the rationale for steering towards rote learning as the most efficacious teaching technique. They were comfortable with it and it imbued all their teaching.

Recitation Scripts

Recitation scripts, Arvind (2015) noted, are a "dominant pedagogic practice... that require convergent factual answers and students' display of known information. It seeks predictable, correct answers" (p. 294). Textbook-based recitation scripts could be verbal or written (Mohan, 2017). During observations, teachers in the study used choral recitation to reinforce new information, reading each sentence as they wrote them on the blackboard, which the class recited back. At the end, each sentence was reread and the class chorused each line. When a new vocabulary word was introduced, it was taught through repeated recitation. Similarly, all lessons ended with teachers writing recitation scripts on the blackboard that captured the main points of the lesson, which the students would spend the rest of the class time copying into their notebooks. This structure of choral and written recitation has been uniformly observed in other studies of LFP schools (Arvind, 2015; Bhattacharya, 2013; Groff, 2017; Sriprakash, 2011).

To ensure students across all sections learned the same information, a head teacher at each grade level in the LFP schools in this study prepared the lesson plans for the grade and determined what the students would be taught. The head teacher of the third-grade explained in an interview, stating, "I am making notes on what I am teaching and next day, other teachers are teaching the same. So (this teacher) is taking the student's notebook to use for her class also."

The primacy of the notes emerges from a context of extremely limited resources, whereby the textbook became the sole point of access to the curriculum for both teachers and students (Boruah, 2017; Mohan, 2017). Given the complexity of introducing academic language to the students in the LFP schools, the teachers' notes became a sort of cheat sheet that whittled down the content to manageable amounts and became the test that students would be assessed on. Therefore, it behooved teachers to ensure their students had their notes in the correct notebooks and intelligibly written. As the next section shows, students who could not comply with these academic expectations were deemed problematic.

Assessing Student Competence and Implications for Academic Failure

According to the teachers, the students they deemed academically successful were those who memorized the notes from each lesson and were able to reproduce them with similarly worded or duplicative questions in the term-end examinations. The teachers set the pace of the class by these successful students; their notes were used as examples and shared with the rest of the class, and they usually occupied the front rows. They were also the students most likely to answer the teachers' questions, often in Hindi and sometimes in English when prompted. In a class of 50 students, there were on average 10 to 15 students in this group, all of whom sat in the front of the room. Teachers often evaluated the effectiveness of the lesson they had taught based on the responses from this group of students. In a post-observation interview, one teacher stated,

> You are seeing, I am asking questions to check, are they getting or not? If they are giving me the correct answer, then I am knowing that they have understood.

Repeated observations of the same teacher, however, found the same group of students raised their hands to answer the teachers' questions and gave accurate responses. These students' copied notes were written neatly and accurately, in "good handwriting," as the teachers said. In informal assessments I conducted during these copying sessions, these students demonstrated their ability to read fairly to very fluently from their notes or textbook. As Boruah (2017) noted, "success in learning or in development of English is equated with the ability to repeat the language of the textbook or teacher's notes, to answer prompt questions, or to copy from the blackboard, and to repeat memorised chunks" (p. 300).

The next group consisted of about 30–35 students who would be the ones protesting when the teacher erased the board, would need to be instructed by their teachers to copy from their top-tier classmates, and were able to do so within class time. These students could answer the teachers' questions in Hindi if also asked in Hindi. Their handwriting was neat and fairly legible. However, in informal assessments I conducted, many were unable to read back what they had written. These students, it seemed, benefitted most from attending the private tuition centers, described in Chapter 6, where the content would be reviewed, providing them with the opportunity to commit it to memory. On one occasion, a teacher walked up when she saw the student stumbling to read aloud to me the words he had just copied from the board. "Don't worry," she said to me, confidently. "In two days, he will know!" I asked her how she knew this. "In class, we will review. Also, he is going to tuition," she replied. Mukhopadhyay and Sriprakash (2011) echoed this reliance on tuition centers and remedial lessons for all students as a way to reinforce learning.

The "Back-Benchers": Language Difficulty or Learning Disability?

The last tier consisted of the remaining five or so students the teachers deemed as being at the bottom of the class, who dutifully copied the notes from their classmates without understanding what they were writing. The teachers used the terms "back-benchers," "slow learners," or "weak" to describe these academically struggling students (Bhattacharya, 2013; Mukhopadhyay & Sriprakash, 2011; Sriprakash, 2011), rather than the formal, medicalized term of "learning disability." They often spoke of students having "low IQ," suggesting they saw these difficulties as an innate cognitive disability, and noted every class had three to four weak students. The significance of being back-benchers was often communicated through their physical placement at the back of the classroom, where the teacher was less likely to walk to and from where the writing on the blackboard was less visible. In Bhattacharya's (2013) study, partial back walls "barely separated [them] from other children in adjoining classrooms" (p. 171).

Like Ladwig and McPherson's (2017) assertion "when 'low-ability' students are identified, it is in terms of what they cannot do—their inability, rather than their ability" (p. 354), the first problematic academic behavior the teachers identified was that the student "was slow in writing." One teacher pointed to the notes a student was working on and said, "See! He is not writing correctly. Just whatever comes in his mind." Students were deemed slow in writing if they took longer copying notes from the board than the rest of the class to the point of not finishing within the class period, or if what they copied was either unintelligible or illegible.

Many teachers identified organizational skills, such as "writing maths in English book" or not "bring[ing] notebooks." One teacher explained, "They are like this only. [He] is always coming without his book. Today he has brought this. Tomorrow he will bring another book." Other problematic academic behaviors included being "weak in maths" because students were "not paying attention." One teacher said a student was "writing but not grasping properly," referring to his difficulty with comprehension, and suggested this was because he had not attended preschool. Of another student, the teacher said, "She understands everything but she's very lazy." The teachers' description of "weak" students as those with "poor handwriting" and "low IQ" illustrated Ladwig and McPherson's (2017) argument that "the presumed links between ability, intelligence and academic tests [are] typically taken-for-granted in schools" (p. 346).

When teachers showed me student work samples comparing the students they deemed as "good" and those they deemed "weak," the differences in handwriting and writing fluency were clear. For instance, in an environmental studies lesson on water conservation that I observed, the teacher wrote five sentences on the board for students to copy as part of a fill-in-the-blanks exercise in the upcoming unit test. Part of Lesson 4, and titled "4. Water for Every Household," the sentences were numbered and read: "1. Machines used for lifting water are called pumps. 2. To ensure good <u>hygiene</u> we must have safe drinking water. 3. <u>Arrangements</u> are made to distribute water all over the town and city. 4. <u>Water</u> should be stored properly and used carefully." One student the teacher had identified as weak wrote the sentences down as:

> 4. Water for every household
>
> 1. Machines used forliing watar called pumps.
> 2. To ens good <u>hygiene</u> we must have ____ safe drinking water.
> 3. <u>Arrangengments</u> are mad eto. distute water all aver The to ity.
> 4. <u>water</u> should be stored properly and used carefuly.

Often, students' struggles with handwriting indicated unfamiliarity even with the letters of the alphabet, slowing down their speed further. In a lesson I observed on English vocabulary of things in a classroom, one student wrote down only 10 words from a list of 20 items in the classroom in the same time as most of the class, although his handwriting was neat. When I asked him, he named most of the actual items in the classroom (table, chair, bag, book, pencil, pencil box, chalk, "color" for crayon, and "rubber" for eraser) in English, plus a few more in Hindi (door, window, duster, water bottle, cupboard), but did not identify the written words when I redirected him to the list.

In informal assessments that I conducted with these students, I found language to be a significant factor in their learning difficulty. For instance, asked by a teacher to reteach the concept of even and odd numbers to a student whom she pointed out as "slow" in mathematics, I used a box of matches as manipulatives. Also, I spoke in Hindi, using the words *doh seh bataa*, or divisible by two, as an informal way to conceptualize the operation, rather than the more formal Hindi academic vocabulary *shum* and *vishum* for even and odd, and used only the English academic word "even" to which the student had been introduced. Thus, I "code-switch[ed] between English and the [student's] dominant language" (de Araujo, 2017, p. 367). After a single explanation with the matches, the student could demonstrate his conceptual understanding of even numbers, but he continued to mix up the word "even" with "equal." Scholars have pointed to the importance of recognizing mathematics as also language-based (de Araujo, 2017), and cognitive thinking being shaped by the language of mathematics (Boroditsky, 2017). It would seem in this instance, the student acquired the mathematical concept of even numbers, but was still confusing the mathematical vocabulary in English. The teacher's response, however, when he consistently used the word "equal" was: "See, he is understanding nothing!" In this way, she dismissed both his conceptual understanding of the math and his linguistic confusion in English. This teacher later informed me she had asked his parents to take their child to "the hospital to get tested," and get a "disability certificate" so he could get help in the exam.

Most of these students sat at the back of the classrooms, some three at a bench. Many were, according to the teachers, "ten rupees" students, who bought "packet snacks" from street vendors, which the teachers did not approve of, rather than bringing home-made snacks. A few did not have uniforms or schoolbooks; the teachers said they were scholarship students who would receive these from the government, which would also pay their fees. When teachers taught a lesson by reading from the textbook and found some students did not have a textbook, they would ask students who did to share with those who did not. However, frequently there would not be enough to go around to the students at the back, and these students would then sit through the class, unable to follow along with the text or the translation. Their participation in the choral recitation was for form's sake; like the students in Mohan's (2017) example who "tried to blend in visually" (p. 20), they joined in the cadence but the words they chanted were unintelligible. The teachers' insistence they write notes only in the notebooks assigned to the particular subject also meant these students frequently sat quietly through the copying sessions without writing notes. To the teachers, these students were "more work" for them. When I asked if having them sit through the lesson rather than writing notes in another book might cause them to fall further behind in class, one teacher said:

> They are giving more work for me. They are not bringing their books, so they are not getting their notes. For them only, I am having Saturday [remedial] class, so we can give the notes.

The "gaze of pathology on children" (Shalaby, 2017, p. xxvii) in schools is well-documented. As Arvind (2015) noted, the "child's academic failure is falsely attributed to her lack of mental abilities or low intelligence and the institutional arrangements go scot-free" (p. 293). Through classroom observations and policy analyses, scholars have documented teachers' construction of students from lower castes as backward in government schools (Mooij, 2008, as cited in Mukhopadhyay & Sriprakash, 2011; Sriprakash, 2011). Bansal's (2018) study of primary schoolteachers' attitudes toward the inclusion of children with disabilities found they were more concerned with the slow learners who "lag behind academically but are not counted among the children with special needs" (p. 119). The process of teasing out the difference between learning and language difficulties in multilingual students is a complex process (Dystefl, 2021; ELT well, 2021; Kormos & Nijakowska, 2017) that, as in Bansal's (2018) and Gupta's (2014) studies, the teachers in this study had received no systematic training on. Further constrained by large class sizes and the English only policy, they "domesticated the pedagogy" (Gupta, 2014, p. 3911), resorting to teaching in ways they were familiar with and were most expeditious for teaching to the test, such as rote learning techniques and teaching through translation (Bhattacharya, 2013; Sriprakash, 2011). Students who could not comply with the teachers' standards for academic success were an additional source of stress for the teachers because they would be most likely to fail. Situated in a fiercely competitive educational environment (Majumdar & Mooij, 2012, p. 207), teachers felt compelled to teach to the test to ensure that students would be successful. As the next section shows, rote learning facilitated this by contributing towards a rote assessment.

Classroom Assessments

In accordance with the requirements of the Continuous and Comprehensive Evaluation (CCE) policy mandated in the Right to Education Act at the time of the study, schools were expected to conduct regular formative[2] assessments of their students at periodic intervals during a term, often called a unit, and summative assessments at the end of each unit (Nawani, 2015, 2015). One purpose behind the frequent assessments was to identify or classify "dull" students so "they could be given requisite support" (Nawani, 2015, p. 35). The Central Board of Secondary Education (CBSE) manual (2010) stated:

> Schools shall diagnose learning difficulties through formative tests right from the beginning of the academic year and bring it to the notice of parents at appropriate intervals of time. (as cited in Nawani, 2015, p. 35)

Although the CBSE (2010) manual cautioned teachers against labeling students as "bright" or "dull," Nawani questioned how "CCE with its system of differential grade allotment would not lead to labeling of students into closed and damaging categories" (p. 36).

Indeed, teachers in the LFP schools often evaluated the effectiveness of the lesson they taught based on the responses from the group of students they perceived as being good students.

The correct answer was frequently, as Nawani (2015) noted, "a memorised response and not a reflection of any deeper understanding on [the student's] part" (p. 36). As a representative of an NGO supporting and promoting English-medium instruction in LFP schools said in an interview:

> The problem is that the assessment system itself is built on rote. The teachers provide remedial help as after-school activities, if they notice that students need that extra support. But when they get to the 10th standard, and they're still struggling, basically we tell the kids, what you need to do is get tested for learning disability and get the certificate, so you can get extended time on the exam and that way you'll pass . . .

The next section illustrates the strategies teachers used to maximize students' chances of passing the test.

Testing Students' Ability to Recall Rather Than Conceptual Understanding

In the LFP schools, systematic class-wide assessments occurred through what they called unit tests, which focused primarily on summative assessments. On average, units were 6 weeks long, including a week to review what teachers referred to as the "portion" or "syllabus," or the curriculum covered, and a week to conduct the test itself. During the review week, students completed practice tests on the portion for the unit test, as a formative assessment to meet the CCE policy requirement. The head teacher for each grade developed the practice test, what she called the "paper pattern," determining what would be on the test and how many marks each question was worth. The practice test was to better facilitate students' performance on the summative assessment, the unit test, which was actual test items taken from the practice test. School and tuition teachers corroborated the same information would be on the test. One teacher said in an interview:

> There may be a small change, such as a different word missing in the fill in the blank question, but the sentence itself will remain the same as on the practice test.

On the last day of review week, the teachers graded and returned the practice tests to the students. Teachers also invited parents to school to copy down the portion to be covered in the unit test, which each teacher had neatly written on the blackboard of their classroom.[3] As the teachers explained to me, this was to ensure the parents would help their children revise for the test.

Almost all the test items themselves assessed recall of English vocabulary rather than students' conceptual comprehension. For instance, a typical test item was fill-in-the-blanks, where the teacher gave the students a sentence relating to the content of the lesson and left out a word in the sentence, which the students were expected to fill in. However, the missing word did not always relate to the academic concept. In a lesson on water storage and distribution systems, one sentence in the test read, "_______ are made to distribute water all over the town or city," where the missing word was "arrangements." In a lesson on animal shelters a test sentence read, "Some animals do not make any ______ to build shelter for themselves," where the missing word was "effort." Thus, students could pass the unit test by memorizing the practice test and reproducing this information in the summative assessment, regardless of their reading ability. Similarly, a national achievement survey conducted by the National Council for Educational Research and Training (NCERT, 2012) found eighth-grade students did better on straight recall questions in science and social science and struggled to answer questions that required application or reasoning. NCERT also found students scored higher answering reading comprehension questions that were passage independent, which were based on prior knowledge (e.g., what color was the banana?) than on passage dependent items, which could only be answered correctly by using information from the text (e.g., what time did Raj leave the office?); as Keenan and Betjemann (2006) pointed out, passage independent items do not necessarily test reading ability or comprehension.

At the end of review week, the teachers pulled out the answer sheets of the students they deemed had done poorly and pointed out to me the students' errors, including poor handwriting, and incorrect spelling and answers. The assumption of only one correct answer meant teachers would reject students' "propositional mis-constructions" (Arvind, 2015, p. 273), even if they were based on linguistic complexity or ambiguity, as illustrated in these examples from one session with a teacher. In grammar, one test item instructed: "Rewrite the sentences using capital letters and full stops [periods]." One sentence was "reena and rohit were surprised." The teacher explained the student's answer "REENA AND ROHIT WERE SUPRISED." would get partial credit because he had included the full stop, but had mistakenly written the entire sentence in capital letters instead of just the proper names. In another instance in math, a student correctly matched number symbols to quantity (i.e., object-numeral representation) and identified the bigger objects from four sets of small and big objects, both test items involving pictorial representation. However, he left blank where he was instructed to "write in words" numbers out of sequence, and "write the missing numbers." When I pointed out perhaps the first student had understood the instruction they use capital letters to mean all capitals, and the second had not understood or been able to read the instructions, she explained: "They know these questions. We review and give many examples in class. This is not new to him. If he doesn't follow the instructions after so many chances..."

As Majumdar and Mooij (2012) noted, the marks race symbolized by the Standard 10 school-leaving exam has become additional gate-keepers for the poor within the educational system that maintains the socioeconomic stratification of the larger society. Although the competitiveness of the educational system inflicts the pressures of school assessment on all students, regardless of socioeconomic status—and, indeed, this is part of the critique—poor students are disadvantaged additionally by their lack of access to linguistic cultural capital (Bourdieu & Passeron, 1990) that hampers their learning. They question "an impoverished notion of education that reduces educational values into mere marks and consequently produces a number of perverse results such as teaching shops, parrot-training, and rote learning" (Majumdar & Mooij, 2012, p. 211). To this, we might add, perceptions of disability.

Conclusion

Access to schooling has not necessarily led to increased quality and better student learning outcomes (Annual State of Education Report [ASER], 2019; Datt & Wang, 2020; Sriprakash, 2011). In 2019, ASER found "a decline in reading ability that correlated with the passage and implementation of the Right to Education Act" (ASER, 2019), while in 2018, the proportion of children in private schools in Std V (Grade 5) who could read Std II (Grade 2) level text at the national level was just 65% (although better than the 44% of children in government schools). Similarly, a study comparing student scores between 2008 and 2012 found as school enrollment grew as a result of the RTE Act, learning deficits rose with a worsening of education deprivation (Datt & Wang, 2020). A National Achievement Survey (NAS) conducted by NCERT (2012) found students from Scheduled Castes, Scheduled Tribes and Other Backward Categories scored significantly lower than students in the General category across all subjects. With quality of education severely compromised, "the conflation of ability with other social or economic markers, such as race and poverty, serves to further stratify students' access to rigorous curriculum and opportunities, and contributes to the replication of future disadvantage" (Parekh, 2017, p. 338). The MOI adds a further layer of complexity when it is English, a language many low-income students do not have everyday access to (Mohan, 2017), so poverty and language are conflated with disability. For instance, the NCERT (2012) study found students performed better on the NAS if the language spoken at home was the same as the language of instruction.

International development and its agenda of EFA has led to the transfer—and translation—of educational structures, policies and practices from the global North to the South, without consideration of goodness of fit (Steiner-Khamsi, 2014), creating new perceptions of in/competence among students (Mukhopadhyay & Sriprakash, 2011). Globalization has placed a premium on acquiring English that disadvantages students whose preferred language is other than English and perpetuates a myth of meritocracy (Phillipson, 1992). Although the task of teaching English and teaching in English is overwhelming for LFP teachers, students who struggle to learn are then perceived to have an intrinsic deficit and deemed incompetent. In studies comparing upper-, middle- and lower-tier, private, English-medium schools, Shenoy and Pearson (2018) found students in lower-tier schools who were acquiring English language skills in school, were twice as likely to be considered at risk for learning disability than students in the high-tier schools. Yet, despite these structural failures of poorly prepared teachers (Saravanabhavan & Saravanabhavan, 2010) and complexities of language (Bhattacharya, 2013; Gupta, 2014), the push for universal access to education and English language skills ignores the exclusion of those at the "bottom of the pyramid" (Vagh & Sharma, 2018). In such situations, as Ladwig and McPherson (2017) stated, "schooling [is] designed to limit learning and legitimate its consequent social exclusions—at least as much as that system is designed to promote learning and social inclusion" (p. 345).

The next chapter explores the limited resources and supports available to teachers in LFP schools, in comparison to those available to students in the upper-tier schools and examines the assessment and evaluation processes that result in students being labeled learning disabled in clinical settings.

Notes

1. As Nawani (2015) noted, "Celebrating individual success stories of poor children studying in challenging circumstances are often highlighted, conveying a message to several disadvantaged children like them that, "if they can do well, why cannot the others"? This conveniently shifts the blame from the system to the individual learner." (p. 39).
2. The meaning of the term "formative assessment" varies in India (Nawani, 2015). In the US, a formative assessment is typically an on-going process of checking for students' understanding in informal ways during the course of a lesson that allows a teacher to make adjustments to their instruction on the spot based on this feedback. Formative assessments can include asking questions either individually or to the whole class through quick polls or app-based quizzes, asking for thumbs up/down to check if students need further clarification, or exit slips at the end of the lesson where students might be asked to write what was most salient in the lesson to them.
3. In several cases, the students copied down the information themselves because the parents could not read or write English. In instances where students struggled to copy the notes, an older sibling was brought in to expedite the process. This system revealed the educational level and competence of each parent and child to everyone in the room.

References

Almeida, R. (2020, June 10). *Tracing the roots of 25 English words with entirely Indian origins.* https://homegrown.co.in/article/44786/tracing-the-roots-of-25-english-words-with-entirely-indian-origins

Anderson, J., & Lightfoot, A. (2018). Translingual practices in English classrooms in India: Current perceptions and future possibilities. *International Journal of Bilingual Education and Bilingualism.* https://doi.org/10.1080/13670050.2018.1548558

Arvind, G. R. (2015). Institutional context, classroom discourse, and children's thinking: Pedagogy re-examined. In M. Thapan (Ed.), *Education and society: Themes, perspectives, practices* (pp. 283–308). Oxford University Press.

ASER (2019). *Annual status of education report, 2018.* Pratham. Anthem Press.

Bansal, S. (2018). Understanding Teachers' Perspective of Inclusive Education for Children With Special Needs (CWSN). *Educational Quest: An International Journal of Education and Applied Social Science, 9*, 1, 115–123.

Bhattacharya, U. (2013) Mediating inequalities: Exploring English-medium instruction in a suburban Indian village school. *Current Issues in Language Planning, 14*(1), 164–184. https://doi.org/10.1080/14664208.2013.791236

Bhatnagar, N., & Das, A. K. (2014). Regular school teachers' concerns and perceived barriers to implement inclusive education in New Delhi, India. *International Journal of Instruction*, *7*(2), 89–102.

Boroditsky, L. (2017). *How language shapes the way we think*. TED Talk. https://www.ted.com/talks/lera_boroditsky_how_language_shapes_the_way_we_think?language=en

Boruah, P. (2017). Learning English in a low-cost semi-urban English-medium school in India: Challenges, interaction patterns and domains of use. In H. Coleman (Ed.), *Multilingualisms and development* (pp. 289–306). British Council.

Bourdieu, P., & Passeron, J. C. (1990). *Reproduction in education, society and culture*. Sage.

Chattopadhay, T., & Roy, M. (2017). *Low-Fee Private Schools in India: The Emerging Fault Lines* (Working Paper 233). National Center for the Study of Privatization in Education. Teachers College, Columbia University. http://ncspe.tc.columbia.edu/center-news/low-fee-private-schools-in-india/

Coleman, H. (2017). Development and multilingualism: An introduction. In H. Coleman (Ed.), *Multilingualisms and development: Selected proceedings of the 11th language and development conference* (pp. 15–34). British Council. http://www.langdevconferences.org/publications/2015-NewDelhi/MultilingualismsandDevelopment-Coleman-ed-completepublications.pdf

Collin, R., & Apple, M. (2010). Globalization, schooling, and literacies in an unsettled age. In S. Tozer, B. P. Gallegos, A. Henry, M. Bushnell Greiner, & P. Groves Price (Eds.), *Handbook of research in the social foundations of education* (pp. 296–307). Routledge.

Cummins, J. (1979). *Cognitive/academic language proficiency, linguistic interdependence, the optimum age question and some other matters* (Working Papers on Bilingualism, No. 19) (pp. 121–129).

Cummins, J. (1981). Empirical and theoretical underpinnings of bilingual education. *Journal of Education*. https://doi.org/10.1177/002205748116300104

Datt, G., & Wang, L. C. (2020). Schooling and learning deficits: A simple unified measurement framework. *Comparative Education Review, 64*(3). http://doi.org/0010-4086/2020/6403-0008

de Araujo, Z. (2017). Connections between secondary mathematics teachers' beliefs and their selection of tasks for English language learners. *Curriculum Inquiry, 47*(4), 363–389. https://doi.org/10.1080/03626784.2017.1368351

DysTEFL2. (2021). *Dyslexia for Teachers of English as a Foreign Language 2.* http://dystefl2.uni.lodz.pl/

ELT well. (2021). *Inclusive language teaching.* http://eltwell.com/resources/inclusive-language-teaching/

EPW Engage. (2021, April 23). *English language education in India: How aspirations for social mobility shape pedagogy.* https://www.epw.in/engage/article/english-language-education-india-aspirations-pedagogy

Frey, N., & Fisher, D. (2009). The release of successful instruction. *Principal Leadership, 9*(6), 18–22.

García, O. (2019). Decolonizing foreign, second, heritage, and first languages: Implications for education. In D. Macedo (Ed.), *Decolonizing foreign language education: The misteaching of English and other colonial languages* (pp. 152–168). Routledge.

García, O., Johnson, S., & Seltzer, K. (2017). *The translanguaging classroom: Leveraging student bilingualism for learning.* Caslon.

Groff, C. (2017). Language and language-in-education planning in multilingual India: A minoritized language perspective. *Language Policy, 16*, 135–164

Gupta, R. (2014). Change in teaching practices: Case of phonics instruction in India. *Procedia: Social and Behavioral Sciences, 116*, 3911–3915. https://doi.org/10.1016/j.sbspro.2014.01.865

Keenan, J. M., & Betjemann, R. S. (2006). Comprehending the gray oral reading test without reading it: Why comprehension tests should not include passage-independent items. *Scientific Studies of Reading, 10*(4), 363–380. https://doi.org/10.1207/s1532799xssr1004_2

Kormos, J., & Nijakowska, J. (2017). Inclusive practices in teaching students with dyslexia: Second language teachers' concerns, attitudes and self-efficacy beliefs on a massive open online learning course. *Teaching and Teacher Education, 68*, 30–41. https://doi.org/10.1016/j.tate.2017.08.005

Ladwig, J. G., & McPherson, A. (2017). The anatomy of ability. *Curriculum Inquiry, 47*(4), 344–362. https://doi.org/10.1080/03626784.2017.1368352

LaDousa, C. (2017). Mind the (language-medium) gap. In K. Kumar (Ed.), *Routledge handbook of education in India, debates, practices, and policies* (pp. 81–96). Routledge. https://www.routledgehandbooks.com/doi/10.4324/9781315107929.ch5

Majumdar, M., & Mooij, J. (2012). The marks race: India's dominant education regime and new segmentation. In C. Sleeter, S. B. Upadhyay, A. K. Mishra, & S. Kumar (Eds.), *School education, pluralism and marginality: Comparative perspectives* (pp. 207–246). Orient Blackswan.

Mili. (2018). Pedagogical reform in Indian school education: Examining the child-centred approach. *Journal of Philosophy of Education, 52*, 3, 533–547.

Mohan, P. (2017). The road to English: The slow migration of the EWS child to elite India. In M. J. Tiwary, S. Kumar, & A.K. Mitra (Eds.), *Dynamics of inclusive classrooms: Social diversity, inequality and school education in India* (pp. 203–215). Orient BlackSwan.

Mukhopadhyay, L. (2020). Translanguaging in primary level ESL classroom in India: An exploratory study. *International Journal of English Language Teaching, 7*(2). https://doi.org/10.5430/ijelt.v7n2p1

Mukhopadhyay, R. & Sriprakash, A. (2011). Global frameworks, local contingencies: Policy translations and education development in India. *Compare: A Journal of Comparative and International Education, 41*(3), 311–326.

Nawani, D. (2015). Rethinking assessments in schools. *Economic & Political Weekly, 50*(3), 37–42.

NCERT. (2012). *What students of Class VIII know and can do: A summary of India's National Achievement Survey, Class VIII, 2012.* Author.

Parekh, G. (2017). The tyranny of "ability." *Curriculum Inquiry, 47*(4), 337–343. https://doi.org/10.1080/03626784.2017.1383755

Phillipson, R. (1992). *Linguistic imperialism.* Oxford University Press

Saravanabhavan, S., & Saravanabhavan, R. C. (2010) Knowledge of learning disability among pre- and in-service teachers in India. *International Journal of Special Education, 25*(3), 132–138.

Shalaby, C. (2017). *Troublemakers: Lessons in freedom from young children in school.* New Press.

Shenoy, S., & Pearson, D. (2018). School culture and its impact on special education practices in Bangalore. *The Journal of the International Association of Special Education, 18*(1), 49–63.

Skutnabb-Kangas, T. (2017). Language rights and bilingual education. In O. García, A. M. Y. Lin & S. May (Eds.), *Bilingual and multilingual education* (3rd ed., pp. 51–64). Springer.

Soler J. (2016). The politics of the teaching of reading. *Prospects, 46*, 423–433 https://doi.org/10.1007/s11125-017-9415-8

Sriprakash, A. (2011). Being a teacher in contexts of change: Education reform and the repositioning of teachers' work in India. *Contemporary Education Dialogue, 8*(1), 5–31.

Steiner-Khamsi, G. (2014). Cross-national policy borrowing: Understanding reception and translation. *Asia Pacific Journal of Education, 34*(2), 153–167. https://doi.org/10.1080/02188791.2013.875649

Tooley, J. (2009). *The beautiful tree: A personal journey into how the world's poorest people are educating themselves.* Penguin.

UNESCO. (2016). *Multilingual education Why is it important? How to implement it?* Author.

Vagh, S. B., & Sharma, D. (2018). Localized assessments can enhance our understanding of learning at the bottom of the pyramid: Insights from India. In D. A. Wagner, S. Wolf, & R. F. Boruch (Eds.), *Learning at the bottom of the pyramid: Science, measurement, and policy in low-income countries* (pp. 121–137). International Institute for Educational Planning.

CHAPTER 6

Assessing for Learning Disabilities

The first chapter presented some ineluctable facts about growing numbers of children being labeled learning disabled. Data from subsequent chapters beg the question: how much of this is related to poverty and students being unfamiliar with the language of instruction, than an innate cognitive deficit that renders them academically unsuccessful on standards set by educational institutions? This chapter examines the differences in classroom resources and external supports that LFP and upper-tier schools have available and identifies how these differences affect both processes and outcomes of labeling and identification of learning disabilities for academically struggling students. It also analyzes some of the more formal standardized assessments administered at learning disability clinics and similar centers, many of which are measurement tools imported from the global North that have been adapted and standardized for Indian student populations. The process of identification is deeply medicalized, with vocabulary, checklists and procedures associated more closely with the United States than "local understandings" (Kliewer & Biklen, 2007) of students' academic struggles and ability (Habib, 2018; Parekh, 2017). Once again, the imbrication of language and learning is overlooked in assessment practices within clinical settings (Karanth, 2008;

M. Kalyanpur, *Development, Education and Learning Disability in India*, Palgrave Studies in Disability and International Development,
https://doi.org/10.1007/978-3-030-83989-5_6

Thapa, 2008) and is imbued with assumptions of universality of knowledge (Serpell, 2000) that disadvantages students who do not have access to this linguistic cultural capital (Bourdieu & Passeron, 1990).

Supports for Academically Struggling Students

In their study comparing upper-, middle- and lower-tier, private, English-medium schools, Shenoy and Pearson (2018) found that lower-tier schools had no resources or supports for the students who were struggling academically and at risk for learning disability, unlike the higher-tier schools which offered special education and resource room services. The Right to Persons with Disabilities (RPWD) Act mandates inclusive government-run and private, government-aided schools provide supports to children with disabilities, including children with learning disabilities, through their SSA programs. The upper-tier, private, unaided schools in this study, which were not required to meet this mandate, charged parents additional fees for specialized services. However, the LFP schools tended to rely more on external sources of support, the "shadow school industry" (Sarangapani & Winch, 2010) of private tuition and coaching that has sprung up to meet this need. As one of the parents in this study put it:

> It's very different if you're rich and have a learning difficulty, you can get everything, all the help you need. If you're poor and you have a learning difficulty, there's nothing.

This section examines these differences in levels of supports as they manifest in schools for students from different socioeconomic backgrounds.

Supports for Academically Struggling Students in Low-Fee-Paying Schools

LFP teachers developed indigenous supports, expecting students to help each other and arranging in-house remedial classes. Most crucially, they depended heavily on the "shadow education" (Ghosh & Bray, 2020) system of private tutoring classes and tuition centers.

Informal In-Class Supports

Teachers felt an acute responsibility to ensure that students copied their notes accurately, given the explicit understanding that having "good notes" would help a student pass examinations by memorizing and duplicating this information in the test. Thus, all the teachers used the class time when students were copying from the board to correct students' notes from the previous day's lessons. When I asked if they could also use some of this time to conduct small group instruction to re-teach concepts for students who may not have understood them in the first attempt, they all responded they had no time. As one of them articulated, "If I do this, when I will mark their answers only?"

Instead, the teachers offered extra support by providing what they called "a little more explanation" on an as-needed basis while they were grading students' notebooks. As one teacher said in an interview,

> I'm calling them separately and asking them, 'how did you get this?' They know the answers, only they are confused. Once I explain again to them, they are understanding.

However, these explanations related more to the mechanics of writing and did not involve re-teaching concepts. For instance, during observations, I noted one teacher told a student she had "not left a line after the question" and another teacher told a student he had "not written the title of the chapter" at the top of the page.

Teachers also expected more academically successful students to help those less so. They instructed students who finished copying the teacher's notes from the board to lend their notebooks to students who were slow writers. One teacher said in an interview she had parents borrow the notebooks from classmates so their child would get the notes they missed, and one parent said in an interview she would contact another student to get the class notes from him for her son. If students needed more individualized supports, the teachers said, the teachers arranged for the students to attend remedial classes.

Formal In-School Supports

School management mandated that teachers provide remedial classes for students whom the teachers felt needed extra help. A common practice among LFP and government schools (Ghosh & Bray, 2020; Mukhopadhyay & Sriprakash, 2011), these classes were typically offered on Saturdays after school. The purpose of the remedial class was to ensure, once again, that students had good notes to study from. As the principal at one of the schools explained:

> The teachers decide who needs the extra class. They are telling the parents their child has to have this extra class because they are not doing well in school. They have to come to the class. Parents don't have to pay, and the teachers themselves are teaching it.

A teacher described this process:

> If (this student) is still not getting, I contact the parents. Explain to them they have to sit with them, help them. We are also having Saturday classes. I have about five weak children. First I am finding them. Then whenever we get time, we call them and review the work. If we cannot complete in this time, then we are calling them for Saturday class.

The review consisted of students copying from a classmate's or the teacher's own notebooks the notes that they had not been able to take down during class hours through the week. I did not observe any additional instruction being provided. Although Ghosh and Bray (2020) found that, in many instances, private secondary school teachers "assume power and coerce" students whom they may be discriminating against to attend supplementary tutoring, parents in this study corroborated the principals' statement that they were not charged for this service and expressed their appreciation for the teachers making time to provide this extra service. Similarly, while Mukhopadhyay & Sriprakash (2011) found that in the rural government schools in their study, *all* the students were selected for the remedial classes for "slow learners" because teachers perceived "all children needing remedial attention in at least one competency" and that children would be "stigmatized if they were singled out for remedial classes" (p. 316), in this study only the students who were struggling in class attended these remedial sessions. It is possible that the imminence of school leaving examinations at the secondary level makes students more vulnerable to pressure from teachers and the effort to reduce the stigmatizing perception of remedial classes in the rural government schools accounts for these differences in the private, urban, primary schools in this study. Instead, the teachers in this study assumed all their students would attend tuition or coaching class outside of school, the "shadow school industry" (Sarangapani & Winch, 2010).

External Supports

With 93.1% of urban students receiving tutoring (Ghosh & Bray, 2020), the expectation that students seek additional support by attending tuition or coaching class was so prevalent that the teachers came to rely on these supplementary services, thus playing "a major role in fostering shadow education" (Ghosh & Bray, 2020, p. 355). The situation was no different in the LFP schools in this study. As a second grade teacher said in an interview,

> Next week we are having unit test. Now we are doing the (practice or review) worksheet, so we are noting where the child is weak and showing the parents. The child is going to tuition so the parents will ask tuition teacher to work on the syllabus.

The need to take tuition classes was unrelated to students' academic performance, but rather an outcome of the poor quality of education in third-tier schools, whether private or government (Ghosh & Bray, 2020; Sarangapani & Winch, 2010). This "shadow cash economy, un-taxed, and under-reported" (Chattopadhay & Roy, 2017) has sprung up as an opportunity for students, coming from average class sizes of 50 and more, to receive somewhat more personalized attention while reviewing in smaller groups the same lessons they have been taught in schools (for a review, see Ghosh & Bray, 2020). The idea of teaching to the test was played out further in the symbiosis between school and tuition. For the top students in the LFP schools, attending tuition classes ensured that they stayed at the top of the class. For the middle tier students, attending tuition classes became a way to learn what was taught in class, be helped to identify what was important to remember, and coached to pass the exam. The fees for private tuition averaged around Rs.400 a month for the lower grades, essentially doubling parents' expenses for their child's schooling (Ghosh & Bray, 2020).

Students who were struggling academically also benefitted from this service. In interviews or during observations, LFP teachers in this study spoke confidently of students "getting" because they went to tuition. Some of the parents who were interviewed also mentioned that their child was going to tuition. One parent asserted that their child's academic problems would go away in time because the child was taking tuition classes.

The tuition teachers and managers interviewed for this study stated that students who were struggling academically would be helped in their centers to get notes from their classmates and to review the exam portion. The lower student-teacher ratio also allowed more personalized support; if a student needed more explanation of a concept, the tuition teacher would provide this. At the secondary school level, however, students had to be *hoshiyaar* (smart) to attend, according to a tuition center manager in this study. As Ghosh and Bray (2020) suggest, since the primary purpose of the coaching classes is to increase the chances of students doing well in the competitive school-leaving examinations, the scores determining future career trajectories, the centers' reputations depended on their clientele's performance. Thus, these services became more discriminating—and more expensive—as students moved up the grades. It is possible that the lower academic expectations at the primary level and the No Detention regulation in the Right to Education Act, which required schools to push all students through the grades until Grade 8 (Nawani, 2015), allowed for more latitude with younger academically struggling students.

These indigenous internal and external supports available in the LFP schools contrast with the imported structures offered in the upper-tier schools, as the next section shows.

Supports for Academically Struggling Students in the Upper-Tier Schools

In contrast, the upper-tier schools had several in school supports for their struggling students, including a team of special education professionals and a learning resource center. As scholars note, these supports are structured very similarly to supports in the US (Ahmad, 2015; Koshy et al., 2017).

In-School Supports

In one school, all students had been assessed on the DIBELS Oral Reading Fluency (ORF) test. Students who had "fallen below range" were referred for follow-up services with the special education teacher. These supports were provided both within the general education classroom and in individual or paired pull-out sessions in a resource center. In the learning resource center, they provided more targeted intervention to students one-on-one or in pairs, structured according to *The Six-Minute*

Solution, a reading fluency program (Adams & Brown, 2007). Supports within the general education classroom were of two kinds. One, the special education teachers offered practical strategies to the general education teachers that they felt "won't burden them more," as one teacher put it in an interview. They noted in interviews that these strategies included changing students' seating, such as bringing them to the front, placing them with a student who was academically more successful, or making sure they could see the board from where they sat. They encouraged teachers to write the lesson objective on the board as well as the new vocabulary that was being introduced and refer to them during and at the end of the lesson. They suggested the increased use of interactive activities, of multi-media such as videos, and games. As the special education teachers explained, many of these strategies were based on the principles of Universal Design for Learning (UDL), a framework for inclusion originating in the global North and mentioned in the Indian Rights of Persons with Disabilities Act (Ministry of Justice, 2016). Two, the special education teachers stated they provided intermittent, on an as-needed basis, or more extended "shadow teacher" service, which would be the rough equivalent of the role of a paraprofessional aide in the United States (Balachandran, August 21, 2014; Sharan, n.d.; "This shadow bridges the gap," 2014),[1] where they spent time providing individualized support to specific students in the general education classroom.

In the other school, the special education teacher stated in an interview that she added modifications to the lesson plans the general education teachers had created, which could include asking students to repeat instructions, using worksheets or the Smartboard as a visual prompt, or providing explanations a second time if it seemed more than half the class had not understood the concept. In addition to working with the teachers, the special education teacher also taught students in smaller groups in the resource center, while the same lesson was being taught simultaneously to the rest of the class in the general education classroom by the general education teacher. These supports are typical of services provided by special education teachers and through learning resource centers at many upper-tier schools and are directly based on structures imported from the US and UK (Ahmad, 2015).

However, not all upper-tier schools provided these services. In these cases, parents were compelled to consider other options for their child, many ending up in the special school in the study. As one parent described:

> He was going to a mainstream school that had absolutely no empathy for his reading challenges and his specific writing issues. He was just being treated as being badly disciplined. They asked me to get him assessed, but even then... Don't get me wrong, I understand that the teachers are under a lot of stress to finish the curriculum, and with the large class sizes, they're doing the best they can.... But I started to ask around, and that's how I heard of (the special school).

The special school, the resource room, the special education teachers and even the efforts at implementing strategies for UDL in the general education classrooms are all structures and practices that have emerged from the global North, and, more significantly, from the field of special education, which has tended to be more exclusionary and segregationist in its focus. The transfer or "borrowing" of these practices (Mukhopadhyay & Sriprakash, 2011), initially through the international development agenda and, more recently, through direct international collaboration between civil society organizations (UNESCO, 2019) has re-created within the Indian context the evolution of special and inclusive education services as it occurred in the global North. Even as inclusive education is being professed in policy and practice, the increasingly narrower parameters of ability have generated perceptions of difference that are gradually leading to the marginalization of students who are "falling below range."

Further, while it would appear that struggling students do have access to services, whether indigenous or imported, the "privatization-by-default" (Ghosh & Bray, 2020, p. 357) and the larger forces of market fundamentalism made most supports available for a price, affecting students at all socioeconomic levels. The modifications for children with learning disabilities listed under the RPWD Act are not many, and private schools are not mandated to provide them. According to the special education teachers in the upper-tier schools, the services they provided were available to any student who was deemed by the general education teachers to need them, and if the parents agreed. In some cases, parents sought to have their child assessed at a private learning disability clinic for a fee and shared the clinic's recommendations for instructional interventions with the school. According to the LFP schoolteachers, if, despite the remedial and tuition classes, students continued to struggle, they could choose to drop out of school or seek to obtain a disability certificate at a free government diagnostic clinic and hospital towards receiving some of the mandated accommodations on the school-leaving examinations in Grade 10. The next section traces this path from school to clinic.

The Process from School to Clinic

Two problems emerge in the identification and labeling process. One is the lack of clear policy guidelines for screening and assessment structures, which results in variations in practice (UNESCO, 2019). The second is the dependence on imported models of assessment, through the use of standardized measures developed in the global North and adapted to Indian populations (Ahmad, 2015; Koshy et al., 2017), and a heavy emphasis on medical practitioners (Ahmad, 2015), including doctors and psychologists, in the assessment process. Most problematic is that the process is imbued with the medical model lens of placing the problem in the child and trying to fix the child, rather than recognizing the social-cultural factors at play (UNESCO, 2019; Velugu, no date, cited in UNESCO, 2019).

Gap Between Policy and Practice in Implementation

Although the RPWD Act urges "detect(ing) specific learning disabilities in children at the earliest," it does not provide guidelines for assessment and issuing of certificate and leaves this to the states (UNESCO, 2019, p. 80). However, as Ahmad (2015) notes, "medical certification is essential for the child having a disability so as to avail the Provisions given by the Education Boards in India, as relaxations during classroom learning and evaluation of the candidate by the Board" (p. 93). According to UNESCO (2019), the Ministry of Social Justice and Empowerment,[2] which has oversight on the education of children with disabilities, provides a process flow chart that indicates a three-step process "starting from screening in school to referral for formal assessment to certification" (UNESCO, 2019, p. 81). The Ministry notification recommends.

> teachers in both public and private schools should carry out screening of children with academic difficulties in Grade 3 or at eight years of age. Every school should have a screening committee headed by the school principal. (UNESCO, 2019, pp. 80–81).

However, according to UNESCO (2019), this notification provides no details to "help schools understand what the screening entails, or refer to sources from where schools can seek information on conducting the screening" (p. 81). As a result, the process from school to evaluation clinic varies by context.

Screening and Referral at LFP Schools

The lack of clear policy guidelines for screening processes allowed teachers at the LFP schools to generate their own structures for screening and parameters for student ability. Their perceptions emerged from the educational context of curriculum pressure, large class sizes, English being the medium of instruction, and the limited training the teachers had received in bilingual education. As Chapter 5 described, based on student performance on the summative assessments that were part of the required Continuous and Comprehensive Evaluation (CCE) policy, students with poor handwriting who had difficulty reading and writing in English, did not complete their notes within the assigned class time or during remedial sessions, and did not recall the pre-determined correct response in the practice or unit tests were perceived as failing. For all the LFP teachers, informing the parents was the first step in this process. In an interview, one teacher described how she expressed her concern to the students' parents:

> See, if the child has a writing problem, if the child is not able to complete all the writing, then I am showing this to the parents. The date is written for when they did the work. Then I am showing them the books of the other students. Let them compare what others have done. On the same day, what the other children have been able to do. Like that I will show. Then they will accept because what the teacher says is true.

Parents were then referred to a Learning Disability Clinic at a government hospital, this teacher continued, although many parents resisted this:

> The slow learners, we are sending them to (the government) Hospital for an IQ checkup. There they are having a learning department where they are testing the students. Or they are giving extra lessons in the (department) just like we are giving extra lessons.

One teacher in the LFP schools acknowledged that language could be a confounding factor in her students' reading difficulties, especially given that English was not a language they were familiar with. As she explained to me:

> Last year, one student is having similar problems. I told her mom to put her in BMC (municipal government) School, Marathi-medium. There she is doing ok. Sometimes the English is confusing him (the current student).

She added that she might recommend to the parents of the current student a similar change in medium of instruction.

Limitations of a Disability Certificate

Although the teachers in the LFP schools spoke about referring parents to the clinic, they were cautious about doing so for at least three reasons that they identified. One, since most parents were resistant to the idea of their child needing a formal evaluation, the teachers felt the need to be judicious about making this recommendation. In an interview, a teacher described:

> Parents are not understanding that siblings are not the same. Some may have high IQs and other has less. We are trying to tell the parents they should take their child to the clinic to get the certificate, but the parents are saying their child is not *paagal* (crazy). Even if we explain that their child is weak and this is not for us but for the child, they are not going. So it is very difficult.

Two, they stated the evaluation was not helpful to them as it provided no instructional recommendations.[3] One teacher at this school informed me a student, a 10-year-old boy, in her Grade 4 class had been referred and assessed the previous year. I asked if the learning certificate had made a difference for the student, and she replied:

> No, I don't understand what the doctor has said. It's the doctor's language only. It's no use. Madam (the headmistress) told (the parents) to get the certificate. She gave them names of hospitals and they have taken him. Now they have sent it to us simply. I don't know what is the use (of it).

Similarly, Velugu (no date, cited in UNESCO, 2019) notes, in her description of a case study of the clinical assessment of a Grade 5 student:

> No school-based intervention for improvement in learning was planned, neither the notes, observations and results of the assessment were shared with the parents or the school. Only the disability certificate was handed over for submission to the school. (p. 81)

The LFP teachers in this study also added that a disability certificate at the primary level was not helpful to the child either. Because the No Retention policy in the Right to Education Act did not allow them to hold or retain a student back, these weak students would just keep getting pushed up to the 8th grade, at which point the modifications for taking the school-leaving exam specified in the law would be made available. Again, Velugu's case study (no date, cited in UNESCO, 2019) corroborates similar thinking:

> After getting the certificate, the school told the parents not to worry as Ridhima will be promoted to the next class until grade 8. In grade 9, her disability certificate will be useful, as it will allow her to choose subjects she finds easy and she will get exemptions in board exams. (p. 81)

Three, when teachers were given a screening tool, they thought the "descriptions (were) too strong," as one teacher stated in an interview, and they hesitated to use it. During this study, a private institute of applied holistic homoeopathy attached to a local BMC children's hospital distributed a screening tool for ADHD (see Fig. 6.1), as a pre-referral step to an assessment in their clinic. As the teacher who showed me the checklist explained, they had been given a week to fill it out on any student they thought met the criteria, and then "someone from BMC would come to pick them up."

The checklist illustrated concerns about the medicalization of students' behavior and the importation of western-centric screening and diagnostic measures (Ahmad, 2015; UNESCO, 2019). The checklist, which was in English, asked for the child's religion along with their name, listing two main domains for ADHD, inattention, and hyperactivity and impulsivity, with academic results and socialization. Teachers were asked to rate these domains on a four point scale of frequency that ranged from "never," "sometimes," "often" and "very often." A small note said "six or more of these" characteristics in each domain indicated that the student could be considered as having ADHD, although it did not clarify at what frequency these behaviors needed to manifest to qualify for this consideration. Nine behaviors were listed under the domain of inattention, six under hyperactivity and three under impulsivity. The student's academic grades and socialization with peers, siblings and elders were to be rated on the same scale. A final section asked for impressions of the student.

ADHD Checklist					
Name:	Age:	Sex:	Religion:	Std.:	DOB:
Symptoms of Inattention* (Six or more of these)					
Does not give close attention to details or makes careless mistakes in schoolwork, work, or other activities.					
Has trouble sustaining attention in tasks or play activities.					
Does not seems to listen when spoken to directly.					
Does not follow through on instructions and fails to finish schoolwork, chores, or duties in the workplace.					
Has trouble organizing tasks or activities.					
Avoids, dislikes, or is reluctant to engage in tasks that require sustained mental efforts (such as homework).					
Loses things necessary for tasks or activities (toys, school, assignments, pencils, books, or tools).					
Is easily distracted by external stimuli.					
Is forgetful in daily activities.					
Symptoms of Hyperactivity and Impulsivity* (Six or more of these)					
Hyperactivity					
Fidgets with hands or feet or squirms in seat.					
Leaves seat in classroom or in other situation in which remaining seated is expected.					
Runs about or climbs when and where it is inappropriate					
Has trouble playing quietly or enjoying leisure activities quietly.					
Is 'on the go' or acts as if 'driven by a motor'.					
Talks excessively.					
Impulsivity					
Burst out answers before questions have been completed.					
Has trouble waiting his or her turn.					
Interrupts or intrudes on others (such as butting into conversations or games).					
Academic Result					
a) Internal Exam grade					
b) Semester Exam grade					
c) Final Exam grade					
d) Writing, spelling error, calculation, reading difficulties					
Socialization					
a) Relationship with peers – Jealousy, possessive, bad habits					
b) Siblings – rivalry, jealousy, quarrelling adjustment					
c) Behavior with elders- Argues, Back answers, refusal of commands					
Evidence of Any Other Mental Disorder					
ASD, ODD, PDD, Schizophrenia, etc.					
Impression:					

Fig. 6.1 This checklist of ADHD behaviors was distributed to the local low-fee-paying schools in the neighborhood by a homeopathy clinic attached to a government hospital (Adapted from an open flyer)

The same clinic also distributed pamphlets in Marathi to schools in the neighborhood, announcing that the first 58 mothers who brought their child to the clinic on a particular Saturday would receive a free checkup and medicines. The clinic would provide homeopathic medicine for children with learning difficulties.[4] When I asked other teachers at this school about the pamphlet and checklist, one teacher informed me she had sent home a pamphlet with one of her student, saying,

> I gave (the pamphlet) to (the student's) parents yesterday, but they have not seen only. I don't care, from my side I will send. If they go to the hospital, they will get free testing.

Another teacher said she felt one student was particularly "mischievous," but that the characteristics described in the checklist were "too strong." According to her, the student did not manifest all these behaviors and certainly not to the point where they needed to be referred to the clinic. Further, she worried that the student's parent would become "very fighting only" with her if she sent in this form. Finally, it turned out, none of the teachers in the school filled out this checklist on their students.

Interviews I conducted with developmental psychologists in government clinics corroborated this process of referral, particularly for secondary students. As one psychologist said:

> Ten years back, schools would not refer students and parents were at the mercy of the school. The tide has now reversed. Schools now tell the parents, but the parent sits on it till the time of the exam. This is because of the social stigma. They come when they have to. This often hurts the students because they don't realize that there can be as much as a six-month wait and there are multiple visits.

According to this clinician, most children came in Class 9 or 10 before they were to take the Class 10 school-leaving exam to get the benefit of the specific provisions for accommodations mandated by the state government.[5] These included extended time, spelling exemptions, an amanuensis (or scribe) or a laptop or a calculator.[6]

Dependence on Imported Models of Assessment

Most of the assessment measures were imported from the US, although a few were adapted or modified for local populations. The upper-tier schools also had the resources to use their own assessment instruments to screen students. Assessments in these schools included the Dynamic Indicators of Basic Early Literacy Skills (DIBELS) and other classroom-based measures used to identify students with learning disabilities. Similarly, the special school conducted a screening and personal evaluation for language skills, social learning and group dynamic skills, using the Childhood

Autism Rating Scale (CARS), to identify the specific services students might need.

The primary assessment measures used in the learning disability clinics were the Wechsler Intelligence Scale for Children (WISC) and the Woodcock-Johnson. According to the developmental psychologists I interviewed, most government clinics used the WISC-3, the older third edition, while most private clinics used the WISC-IV, its more recent version. The WISC is an IQ or an aptitude test that is assumed to measure a student's potential while the Woodcock-Johnson is an achievement test that is assumed to measure the level the student is currently performing at. This difference between these two scores provides "a discrepancy factor based on achievement and cognitive scores," a reference point that is used both in the United States and in India to determine the presence of a learning disability (Ahmad, 2015; Unni, 2012). In addition, the clinicians explained in interviews, clinics used the TOWL (a test of written language), CTOPP (a test for phonemic awareness), the Gray Oral Reading test (for oral reading, comprehension). Some clinics also used the Wide Range Achievement Test (WRAT). A psychologist at a private clinic in an interview clarified that sections of the Woodcock-Johnson, including "fluency, reading, spelling, phonological awareness and comprehension, (and) some maths," were used on a regular basis, and the entire test was used only for students who needed accommodations to take the International Baccalaureate (IB) and IGCSE school-leaving exams.

Some assessments are popular because they do not have verbal or language-based components, such as the Raven's Matrices test. Similarly, the Seguin Form Board Test (SFBT), which was developed in 1856 and measures visual-motor coordination only as a proxy for intelligence, is also used in some contexts as a "culture-fair cognitive test" (Koshy et al., 2017, p. 276).

Problems in Using Imported Assessments

These assessment measures are problematic in some ways. All of them are imported from the US, although some have been translated into local languages (Ahmad, 2015; Koshy et al., 2017; UNESCO, 2019). For instance, the WISC-3 is available in Hindi and Gujarati. However, very few, if any, had been standardized to Indian norms.[7] The WISC-IV was one of the few standardized for use in India, although this standardization appeared minimal. According to the booklet, the Indian version of the WISC-IV (2012) is normed on just 334 children (118 boys, 153 girls) with almost 84% parental educational levels of college graduate and above, studying in schools with English as medium of instruction. The booklet also mentions, in the piloting, children were shown a picture of a woman's face and asked to identify what was missing. So many of them pointed out her bindi, the vermillion dot worn on Indian women's foreheads, was missing, rather than the eyebrow, that the test was "modified" to make sure the final picture included a bindi on the woman's forehead. Yet, students tested with this same assessment measure are being asked to identify a snowman and a sled, and answer questions like "when did the pony express end?" and "before pioneers settled throughout North America, what one group of people was credited with leading the westward movement?" They are also asked to name four seasons, including spring and autumn; although monsoon is one choice, the clear delineation of four seasons as occurs in temperate climate does not occur in tropical India, and changes in weather tend to be associated more with spring or harvest festivals than with seasons (Boruah, in press).

Even tests like the Seguin Form Board, which is seen as "an effective community-based intelligence-testing tool in resource-limited settings" (Koshy et al., 2017, p. 279), have been found to have very different results depending upon the population. Researchers have recommended establishing "separate norms to account for differences in age, sex, socioeconomic level, and residential setting (rural/urban)" (Koshy et al., 2017, p. 279). Similarly, Sinha (1968), noting that engineering and architecture students in India tended to score better on the Raven's Matrices test due to its primary focus on spatial and psychomotor skills, recommended the need for a study of the validity of its use. However, the test continues to be used as a proxy for intelligence within the medical field (Dhanu et al., 2019).

Strong Reliance on Medical Practitioners

The heavy dependence on the medical practitioners and psychologists for the assessment process also follows the avenues established in the global North. Although clinics are permitted by the state government to conduct the diagnostic assessment, only specific state government-run hospitals are permitted to provide the actual disability certificate; municipal hospitals and private clinics are allowed only to provide an "opinion certificate" or an evaluation report (UNESCO, 2019). For instance, in Maharashtra, "the state board recognizes certification only from three hospitals – Lokmanya Tilak Municipal Medical College, Sion, Mumbai (Sion Hospital) and Nair Hospital; KEM Hospital, Mumbai also certifies children having SLD or from the Government Surgeon General in Mumbai" (Sandhu, 2015). The professionals at the state and municipal hospitals and diagnostic clinics include both medical personnel, such as developmental pediatricians, and educational personnel, such as developmental psychologists and special educators, whereas most private clinics had primarily educational personnel (Ahmad, 2015; UNESCO, 2019). In separate interviews with developmental psychologists in private and government clinics, I was told that private clinics typically charged between Rs. 2,500 and Rs. 6,000, which made affluent parents their primary clientele, whereas most of the clientele at the government clinic attended SSC (or state government school-leaving exam) schools, indicating more middle- and lower-socioeconomic status clientele.

The disability certificate itself was 2 pages long. The first page was a cover letter, addressed to the school principal, specifying the team members who had conducted the assessment, such as Developmental Pediatrician, Clinical Psychologist, Occupational Therapist, Psychiatrist and Special Educator and the diagnosis. The second page was the actual certificate, for instance, for learning disability. After stating the date, the child's name, their date of birth, sex, father's and mother's names and school name and date of registration as LD, it provided spaces for details on visual and auditory acuity, to indicate that the possibility of the student's hearing or vision being a factor for the academic difficulty had been ruled out. There were additional small spaces for information on the physical and neurological assessment, the psychological assessment, including "full scale IQ," and educational assessment. A section called "Interpretation" gives three choices of dyslexia (वाचनातील अध्ययनअकार्यक्षमता *vāchnāteel addhyayan akāryakshamatā*), dysgraphia (लेखनातीलअध्ययन अकार्यक्षमता *lekhnāteel addhyayan akāryakshamatā*) and/or dyscalculia (अंकगणितातील अध्ययन अकार्यक्षमता *ankganitāteel addhyayan akāryakshamatā*), with a diagnosis of learning disability (अध्ययन अक्षमता *addhyayan akshamatā*). The last section asked for recommendations in two domains, remedial education and/or provisions for LD.

Locally Developed Screening and Assessment Instruments

Efforts have been made to develop assessment measures more appropriate for local populations (Karanth, 2008; Thapa, 2008). Many have focused on providing translated versions of the originals. In other cases, the local context for normative academic and behavioral expectations has been taken into account. For instance, one of the best-known NGOs in education in India has developed a screening tool that is currently being used for children with disabilities as part of the child find programs. Constructed very much on the lines of the yes/no questions of the Ten Question Survey Instrument (TQSI) used in prevalence studies (Croft, 2013), the questionnaire, available in Hindi[8] and Marathi, asks 16 questions relating to the "Identification of Special Needs for School-Going Children" for learning disabilities. Some questions that pertain to behavior and executive functioning ask if the child is able to work without being distracted or complete the task if the caretaker is not around, is able to focus for a limited amount of time while working in the school or home or while playing with neighborhood children, or finds it difficult to learn rules while playing with similar aged kids. Other more academic-based questions ask if the child forgets to read any word when asked to read a sentence from a paragraph, or if the child finds it difficult to understand any word from a sentence or answering any question from a paragraph or a story. Questions on writing ask if the child has difficulty taking notes from the blackboard into his/her notebook, maintaining distance between two words or lines, or using full stop while reading and writing. Getting more specific, it asks if the child has difficulty comprehending certain mathematical symbols or relationships of numbers or identifying numbers or letters that look similar, getting confused between some words or numbers such as "b" and "d," "p" and "q," "u" and "n," or 9 and 6. Almost as though it is assumed that English would be the cause of the child's problems, these examples of letters and numbers are in English. It ends with the note that, if the survey-taker received affirmative responses to 4 or 5 questions, then it is recommended that the child be taken for a consultation with an appropriate expert. Similarly, the Annual State of Education Report (ASER) tool has been modified for children with disabilities with support from Ministry of Social Justice and Empowerment and piloted with 50 schools on learning outcomes of numeracy and language and functional ability (Pratham, 2019).

In response to the need for more culturally contextualized assessments, the National Brain Research Council developed the Dyslexia Assessment in Languages of India or DALI as the first nationally standardized screening tool for learning disabilities (Rao et al., 2021). As Dr. Nandini Singh, who headed the development of this tool, said in an interview:

> The DALI wasn't just a translation. It was culturally adapted and translated by native speakers of the language who had an understanding of the writing system, and then piloted with 10,000 children. We asked teachers to use it for every child in their classroom.... as a universal screening (which helped to account for teacher bias and to establish benchmarks). We found that the screening tool is a good predictor of reading problems.... Most important, what we noticed was that despite growing levels of exposure to English, students for whom the spoken language is not English do much better in the Indian languages. We then assessed those who did poorly in both. We felt this established some sort of neurological basis to the reading difficulty.

Another indigenous alternative for screening for learning disabilities is the Diagnostic Test of Learning Disability (DTLD) developed in 1993 by Smriti Swarup and Dharmishta Mehta at the SNDT Women's University. Constructed "to identify those children who experience learning problems because of learning disability" it assesses six areas in visual and auditory perception: eye-hand coordination, figure-ground perception, figure constancy, position in space, spatial relations and auditory perception based on the multisensory approach of visual-auditory-kinesthetic-tactile or VAKT of Marianne Frostig in 1964 and Grace Fernald in 1943. It also assesses four aspects of cognitive functioning, such as memory, cognitive abilities, receptive language and expressive language. It was piloted with 250 "poor achievers" in Class 4 and finally administered on 1050 children for reliability and validity towards standardization. The latter group were "within the age range of 8-11 years drawn from 17 private aided English-medium schools of Bombay. Institutional variables like number of students, medium of instruction, economic status of the school, teacher qualifications were taken into consideration while selecting the sample. All the students who were identified as low achievers in a class constituted the sample" (Swarup & Mehta, 1993, p. 10). However, no details on these variables or the basis on which students were deemed as low achievers are provided. This assessment is included in the pre-service teacher training curriculum for the RCI certified Bachelor's degree program in special/inclusive education.

A third assessment tool is the Linguistic Profile Test (LPT) developed by Pratibha Karanth in 1980 and modified in 1984, originally in Kannada and now available in Hindi, Malayalam, Telugu and Tamil. The Test "evaluates language at the phonological, syntactic, and semantic levels through a wide range of tasks such as pointing, naming, repetition, indication of grammatical acceptability, rapidly listing words from given lexical categories, sentence completion, and matching synonyms and antonyms" (Karanth, 2008).

Problems with Equating Reading and Writing Ability with Intelligence

A UNESCO (2019) report notes that "lack of assessment in schools can lead to high referrals for specialized assessment of learning disabilities (LDs), and inappropriate formal assessments can lead to inaccurate diagnosis, inflating the incidence of LD" (p. 13). Yet, locally developed alternative diagnostic or screening tools are either not being used, particularly in private clinics which prefer the clout of an evaluation that involved American measures, or include assessment domains that have since become outdated. Many of the clinicians I interviewed expressed concern about the lack of a systematic process and appropriate tools for screening for dyslexia, especially given the rising numbers of children who seemed to be requiring this service. As a clinician stated in an interview:

> ASER was providing this disturbing information that so many children were not reading at grade level. The numbers for dyslexia were rapidly increasing. Yet there was no systematic process in place for diagnosis or even the appropriate tools for any kind of assessment.

However, the concern is not so much that there is no systematic process for diagnosis or the appropriate tools for assessment, but that the clinicians appeared to see a correlation between children not reading at grade level and an increase in the number of children being assessed for dyslexia. Although children not reading at grade level may have much more to do with poor reading instruction and less with their having dyslexia and being unable to learn to read, the conclusion was that difficulty in reading is because of an intrinsic deficit in the child. While some teachers and professionals acknowledged that medium of instruction or language may play a large role in children having problems reading, most felt that assessment measures which accounted for this language difference by evaluating children in their preferred language as well as English were adequate and could identify what they saw as a cognitive deficit.

In their study of local understandings of literacy in people with severe disabilities, Kliewer and Biklen (2007) describe how, in attempting to create a nation of readers, phonics instruction came to be understood as a precursor to reading. As they state:

> Phonemic awareness in early or initial literacy is not unreasonable. That it has become standardized as the sole focus of resources, however, is not in the best interests of a literate citizenry.... The medical model reduces literacy instruction to an *intervention* akin to ingesting a pill. (p. 2586, emphasis in original)

The international agenda of Education For All, while providing access to education for millions who were denied, has also contributed to this "global norming" (McDermott et al., 2011), of universal standards of literacy, competence and ability (Le Fanu, 2013, Yates, 2018). By requiring that all children be able to read, students who do not are now perceived as deficient, somehow learning disabled, labeled and marginalized.

The complexity of assessments also raises deep epistemological questions on what is valued knowledge (Grosfoguel, 2013). Does a fill-in-the-blanks test item like "Some animals do not make any ______ to build shelter for themselves," where the correct expected response was *effort,* as we saw earlier, truly assess a student's knowledge of animal shelters? "Can any attempt to measure intelligence predict a person's value or potential to contribute meaningfully to the world?" asks Chris Cooper, the parent of a disabled young man (Habib, 2017). Does everyone need to learn algebra (Black, 2011), and why is indigenous Baiga *vidya* or knowledge about the medicinal qualities of plants less valued (Sarangapani, 2003)? The assumption of universality of knowledge and of what is valued knowledge impacts both school curriculum and assessment measures as avenues towards establishing parameters of student competence towards the ultimate aim of a global labor force.

These arbitrary parameters of student ability (Parekh, 2017) have also led to the ubiquitous assumption of IQ as a measure of intelligence and the adoption of western-centric assessment tools, despite extensive research identifying IQ as a cultural construct (e.g., Serpell, 2000). As Serpell's iconic study showed, constructs such as pattern reproduction skills, which are the mainstay of IQ and other assessments, are culturally defined: students in Zambia had difficulty drawing a car because of their limited exposure to paper and pencil tasks, and students in Britain drew adeptly. On the other hand, the Zambian students excelled in creating scaled, three-dimensional models of cars using a single length of wire, and the British students failed the task. Like my student in the United States who knew the word "*jeopardy*" because he saw it every day on the television show but not the word "*floor*," the assumptions that, for instance, vocabulary can be de-contextualized, and the vocabulary of a student in India, even in English, will be the same as that of a student in the United States is troubling. These students who are being labeled learning disabled are me, when I was learning Khmer in Cambodia at the age of 50 for my work. The differences in letters in this ornate script were indistinguishable to me, so I could not learn to read, and had to focus on oral fluency. But my short-term memory was not good enough to retain sufficient words to build a speaking vocabulary. I would write elaborate sentences in Devanagari script, using the vocabulary I did have, only to be unable to understand the meaning of what I had written myself when I went back to read them. These students are me, when I was learning Hindi, the national language, in primary school in India. Hindi was the fourth language I was learning in my life, and doing so primarily in a school setting. Although I learnt the script and the vocabulary, I never gained the oral fluency and capacity of a native speaker. The difference between me and these students, however, was that I was privileged to be fluent in a language, English, that had more global and national clout than Hindi or Khmer, and so I was allowed to become academically successful and still deemed professionally competent.

Notes

1. Shadow teachers, who charge Rs. 5000/- to Rs.10,000/- per month, have become increasingly popular as a special education option for parents who can afford to pay for this service (Sharan, n.d).
2. As per a notification dated January 4,2018 by Department of Education of Persons with Disability, Ministry of Social Justice and Empowerment (UNESCO, 2019).
3. On the other hand, both middle-class parents and developmental psychologists at private clinics stated that the assessment reports they received or provided respectively were extensive, sometimes running to 10 pages and included detailed recommendations for instructional or classroom interventions.

4. In her case study on the diagnostic assessment of a Grade 5 student, Velugu (no date, cited in UNESCO, 2019) stated that medication, particularly allopathic, is becoming increasingly popular as a response to treat ADHD. This was corroborated by a psychologist at a learning disability clinic in an interview.
5. On the other hand, the psychologist at the private clinic informed me that many of the more affluent parents were hoping to send their children to study abroad and came to have their children assessed so they could avail of accommodations at universities in the United States and UK. For this report, universities demanded that students underwent a full battery of internationally recognized and standardized tests. However, this could not be corroborated further.
6. According to Sandhu (2015), "the provisions provided by the Maharashtra State board of Secondary and Higher Secondary Education are as follows: 25% Extra time, Oral Test along with the written examination for Standard I to IX, Promotion to next class on the basis of the Average, Writer/ typewriter, Question papers to be read out, Exemption from the Second and Third Language, Physiology-Hygiene and Home Science in place of Science, Arithmetic Standard I to IV—errors of interchanging of computation signs and number reversal to be ignored, Arithmetic of Standard V in Standard V & VI, Arithmetic of Standard VI in Standard VII & VIII, Arithmetic of Standard VII in Standard IX & X, Use of Calculator at Class X Examination, Spelling errors and incorrect sentence construction to be ignored, Errors in respect of showing directions in Geography to be ignored, Students between Standard I & IX exempted from drawing diagrams, graphs and charts, Marks for such questions proportionately distributed to the other questions or students to be provided with supplementary questions, 20% consolidated grace marks in one" (p. 86).
7. Although the Indian Adaptation of the Stanford Binet Test of Intelligence by S.P. Kulshresta (Gopalkrishnan & Venkatesan, 2019), also known as the Binet-Kamath's Test of Intelligence, does exist and is one of the assessments included in the curriculum framework for the undergraduate degree program for special/inclusive education developed by the Rehabilitation Council of India, it was not mentioned as one of the assessments currently being used in clinics, perhaps because newer alternatives are available.
8. In keeping with Mohan's (2017) distinction between Hindi, as the more formal academic language, and Hindustani, as the more informal language of the streets, the questions are largely stated in

Hindustani. For instance, it uses more familiar and commonly used words like "*tukleef*" rather than "*kutthinayin*" and "*sawaal*" rather than "*prushn.*".

References

Ahmad, F. K. (2015). Exploring the invisible: Issues in identification and assessment of students with learning disabilities in India. *Transcience: A Journal of Global Studies, 6*(1), 91–107.

Adams, G. N., & Brown, S. M. (2007). *The six-minute solution: A reading fluency program.* Sopris West.

Balachandran, K. (2014, August 21). *Turning the spotlight on the shadow teacher.* Deccan Herald. https://www.deccanherald.com/content/426628/turning-spotlight-shadow-teacher.html

Black, C. (Director). (2011). *Schooling the world: The white man's last burden* [Film]. Lost People Films. https://schoolingtheworld.org/film/

Bourdieu, P., & Passeron, J. C. (1990). *Reproduction in education, society and culture.* Sage.

Chattopadhay, T., & Roy, M. (2017). *Low-Fee Private Schools in India: The Emerging Fault Lines* (Working Paper 233). National Center for the Study of Privatization in Education. Teachers College, Columbia University. http://ncspe.tc.columbia.edu/center-news/low-fee-private-schools-in-india/

Croft, A. (2013). Promoting access to education for disabled children in low-income countries: Do we need to know how many disabled children there are? *International Journal of Educational Development, 33*, 233–243.

Dhanu, G., Havale, R., Shrutha, S. P., Quazi, N., Shafna, T. P., & Ahmed, A. (2019). Assessment of Intelligence Quotient using Raven's Coloured Progressive Matrices among school children of Hyderabad Karnataka region and its correlation with prevalence of dental caries. *Journal of the Indian Society of Pedodontics and Preventive Dentistry, 37*(1), 25–30. https://doi.org/10.4103/JISPPD.JISPPD_236_18

Ghosh, P., & Bray, M. (2020). School systems as breeding grounds for shadow education: Factors contributing to private supplementary tutoring in West Bengal, India. *European Journal of Education, 55*, 342–360. https://doi.org/10.1111/ejed.12412

Gopalkrishnan, I. K. & Venkatesan, S. (2019). Normative congruence between 1967 and 2002 adaptations of age scale for Indian urban children. *The International Journal of Indian Psychology, 7*, 579–590. https://doi.org/10.25215/0704.067.

Grosfoguel, R. (2013). The structure of knowledge in westernized universities epistemic racism/sexism and the four genocides/epistemicides of the long 16th Century. *Human Architecture: Journal of the Sociology of Self-Knowledge, 11*(1), 73–90.

Habib, D. (2017). *Intelligent lives.* New Hampshire Charitable Foundation. Trailer: https://www.youtube.com/watch?v=jjyENb2qbu0

Habib, D. (Director). (2018). *Intelligent lives.* [Film]. UNH Institute on Disability. https://intelligentlives.org/

Karanth, P. (2008). Learning disability and language learning. In K. Thapa, G. M. van der Aalsvoort, & J. Pandey (Eds.), *Perspectives on learning disabilities in India: Current practices and prospects* (pp. 80–96). Sage.

Kliewer, C., & Biklen, D. (2007). Enacting literacy: Local understanding, significant disability, and a new frame for educational opportunity. *Teachers College Record, 109*(12), 2579–2600.

Koshy, B., Thomas, H. M. T, Samuel, P., Sarkar, R., Kendall, S., & Kang, G. (2017). Seguin Form Board as an intelligence tool for young children in an Indian urban slum. *Family Medicine and Community Health,* 5(4), 275–281. https://doi.org/10.15212/FMCH.2017.0118

Le Fanu, G. (2013). Reconceptualising inclusive education in international development. In L. Tikly & A. Barrett A (Eds.), *Education, quality and social justice in the global south* (pp. 40–55). Routledge.

McDermott, R., Edgar, B., & Scarloss, B. (2011). Global norming. In A. Artiles, E. Kozleski, & F. Waitoller (Eds.), *Inclusive Education in five continents* (pp. 223–235). Harvard Education Press.

Ministry of Law and Justice. (2016). *The Rights of Persons with Disabilities Act.* Legislative Department. http://legislative.gov.in/actsofparliamentfromtheyear/rights-persons-disabilities-act-2016

Mohan, P. (2017). The road to English: The slow migration of the EWS child to elite India. In M. J. Tiwary, S. Kumar, & A. K. Mitra (Eds.), *Dynamics of inclusive classrooms: Social diversity, inequality and school education in India* (pp. 203–215). Orient BlackSwan

Mukhopadhyay, R. & Sriprakash, A. (2011). Global frameworks, local contingencies: policy translations and education development in India. *Compare, 41,* 311–326. https://doi.org/10.1080/03057925.2010.534668 http://www.informaworld.com

Nawani, D. (2015). Rethinking assessments in schools. *Economic & Political Weekly, 50*(3), 37–42.

Parekh, G. (2017). The tyranny of "ability". *Curriculum Inquiry, 47*(4), 337–343. https://doi.org/10.1080/03626784.2017.1383755

Pratham. (2019). *Assessment for All: An inclusive assessment tool.* http://assessmentforall.org/

Rao, C., Sumathi, T. A., Midha, R., Oberoi, G., Kar, B., Khan, M., Vaidya, K., Midya, V., Raman, N., Gajre, M. & Chatterjee Singh, N. (2021). Development and standardization of the DALI-DAB (dyslexia assessment for languages of India – dyslexia assessment battery). *Annals of Dyslexia*. https://doi.org/10.1007/s11881-021-00227-z

Sandhu, P. (2015). Legislation and the current provisions for Specific Learning Disability (SLD) in India: Some observations. *Journal of Disability Studies, 1*(2), 76–79.

Sarangapani, P. M. (2003). Indigenising curriculum: Questions posed by Baiga *vidya*. *Comparative Education, 39*(2), 199–209.

Sarangapani, P. M., & Winch, C. (2010). Tooley, Dixon and Gomathi on private education in Hyderabad: A reply. *Oxford Review of Education, 36*(4), 499–515. https://doi.org/10.1080/03054985.2010.495465

Serpell, R. (2000). Intelligence and culture. In R. J. Sternberg (Ed.), *Handbook of intelligence* (pp. 549–577). Cambridge University Press. https://doi.org/10.1017/CBO9780511807947.026

Shenoy, S., & Pearson, P. D. (2018). School culture and its impact on special education practices in Bangalore, India. *The Journal of the International Association of Special Education, 18*(1), 49–63.

Sinha, U. (1968). The use of Raven's Progressive Matrices test in India. *Indian Educational Review, 3*(1), 75–88.

Swarup, S., & Mehta, D. (1993). *Diagnostic test of learning disability*. Department of Special Education. S.N.D.T. Women's University, Bombay, India

Thapa, K. (2008). Learning disabilities: Issues and concerns. In K. Thapa, G. M. van der Aalsvoort, & J. Pandey (Eds.), *Perspectives on learning disabilities in India: Current practices and prospects* (pp. 23–47). Sage.

UNESCO. (2019). *State of the education report for India 2019: Children with disabilities*. Author.

Unni, J. C. (2012). Specific learning disability and the amended "Persons with Disability Act". *Indian Pediatrics, 49*, 445–447. http://medind.nic.in/ibv/t12/i6/ibvt12i6p445.pdf

Yates, R. B. (2018). *Aid, attitudes and ambivalence: The impact of Australian aid in the introduction and sustainability of inclusive education initiatives in the Republic of Kiribati* (Unpublished PhD dissertation). Queensland University of Technology

CHAPTER 7

Conclusion

Research indicates that language barriers often become factors for wrongful identification of learning disabilities among emerging English language learners (Garcia & Wei, 2014; Klingner & Artiles, 2003), a situation that is especially relevant within the multilingual Indian context. In India, an education in English, a colonial legacy, has been the bastion of the elite for several decades, primarily because of the high costs of a private education (Annamalai, 2004; Jhingran, 2017; Mohan, 2017). Government schools have traditionally offered a free education in the regional or national language. However, the realization that the acquisition of English offers access to elite higher education institutions as a pathway to social mobility has been growing, and more middle-class parents have been choosing to send their children to high-fee-paying or upper-tier, English-medium, private schools (Kumar, 2017; LaDousa, 2017).

Since the economic liberalization in the 1990s, low-fee-paying private schools catering to low-income families have burgeoned (Tooley, 2009). Although they are fee-based, they are attractive to parents because many offer English as a medium of instruction, giving low-income children access to a much sought-after skill. As a result, increasing numbers of low-income children who do not have everyday access to the language are choosing to study in low-fee-paying, private, English-medium schools (Chattopadhya & Roy, 2017; Juneja, 2017; LaDousa, 2017). In 2013,

M. Kalyanpur, *Development, Education and Learning Disability in India*, Palgrave Studies in Disability and International Development,
https://doi.org/10.1007/978-3-030-83989-5_7

government statistics showed that private schools enrolled 33% of the total student population while constituting 22% of total schools nationally (NUEPA, 2013, as cited in Chattopadhya & Roy, 2017). The 2019 Annual State of Education Report (ASER) noted that between 2006 and 2014 private school enrollment increased steadily from 18.7 to 30.8%. Traditional social hierarchies of caste/class, which have been and continue to be the bane of Indian society are slowly being upended in the process (Dutt, 2019; Yengde, 2019). Yet, inequities persist, particularly within the realm of education (Vulli, 2014), with English becoming the new caste marker in society (Bhatia, 2018).

As part of the Education For All (EFA) campaign and the impetus toward universal education, more children than ever before are now in school. However, India's masses place an immense strain on an educational system, unable to meet the demand for schools at any level, least of all English-medium, resulting in a fiercely competitive educational environment. With this educational bottleneck, a new disability category, that of learning disability, has begun to emerge (Karanth, 2008; Ramaa, 2000). The 2011 Census found that children with learning disabilities constituted 7% of all children with disabilities between the ages of 6 to 14 (UNESCO, 2019), while other sources place prevalence rates at 14% (see Panshikar, 2019). In 2007, the Bollywood movie, *Taare Zameen Par,* which sought to create mainstream awareness of dyslexia, was a blockbuster success. In 2009, the category of specific learning disabilities was formally introduced with the amendment of the 1995 Persons with Disabilities Act, now the 2016 Right of Persons with Disabilities Act. Although it currently has the lowest incidence of all disability types in India (UNESCO, 2019), its evolution strongly parallels the way this category had evolved in the US, where it constitutes 50% of all children with disabilities. In the US, too, the category emerged from a competitive schooling environment during the Space Race in the 1950s, resulting in academically struggling students becoming marginalized and labeled (Sleeter, 1986). Over time, students from low-income and linguistically diverse backgrounds have come to be over-represented in this category (Duhaney, 2014), just as is occurring in India. The similarities in this trajectory are disconcerting. It would seem in the process of educating everyone, India may also be taking on some of the unexpected and unfortunate consequences of becoming a developed country.

A main indicator for development on the global stage is an educated citizenry and the provision of universal access to education. Although this has led to the expectation that all children learn to read, this outcome is not guaranteed for low-income students attending poor quality schools (ASER, 2019). The additional expectation that students learn English, if they aspire toward social mobility, places students from low-income and linguistically diverse backgrounds at a further disadvantage. Low-fee-paying (LFP) private English-medium schools have been regarded as an appropriate and effective response to providing access to quality education to low-income students (Day et al., 2014; Heyneman & Stern, 2014; Tooley, 2009), given that many governments in the global South have been unable to meet this need. However, there is concern that by being fee-based, such schools continue to exclude the poorest of the poor (Chattopadhya & Roy, 2017). On the other hand, to be considered low-fee, the schools are required by law to keep their tuition rates below a specified amount. To cut costs, then, schools often hire poorly trained teachers and maintain minimal curricular resources (Bangay & Latham, 2013), which results in compromised quality and has led to the emergence of a supplemental "shadow education" system of unregulated, fee-based private tuition centers and classes (Ghosh & Bray, 2020; Sarangapani & Winch, 2010). The immersion model of instruction that the schools follow adheres to an "English only" policy that results in students' lack of academic proficiency in English and a subtractive linguicism with the loss of their own preferred language (Shenoy & Wagner, 2019). Additionally, few teachers have received systematic pedagogical training relating to bilingual education (Bhattacharya, 2013; Gupta, 2014).

What the Study Found

Using qualitative research, this book explored why the number of students being labeled learning disabled has been increasing since the 2010s, and if learning in English could be affecting students' ability to learn. A document analysis of legislative policies situated the emergence of the label of learning disability within the larger sociopolitical context of the international aid agenda of EFA by 2015 and national economic reforms liberalizing access to private education. Through classroom observations and interviews with school personnel and parents, and psychologists at assessment clinics, the study focused on low-income students with little access to English other than by studying at private, low-fee-paying English-medium schools. To offset the impact of language on learning, the study also examined the experiences of affluent and middle-class students fluent in English who were struggling academically in private, upper-tier English-medium schools. A private, special school for students with learning disabilities was also included in the study.

English as a Factor in Maintaining Societal Hierarchies of Caste/Class in Schools

LFP private schools have government-mandated caps on tuition to be considered low-fee-paying (Chattopadhay & Roy, 2017). In keeping with the global neoliberal ideology of productivity and efficiency, LFP schools tend to cut costs by increasing class sizes and employing less qualified teachers at lower salaries (Bangay & Latham, 2013), while international development aid's gender mainstreaming policies have led to women predominating in these educational spaces (Manjrekar, 2013). Further, with more low-income students accessing English-medium education, socioeconomic hierarchies in India are played out in the subtleties of quality of English, such as "convent-educated" versus "regional" accents (Bhatia, 2018; Deb, 2011).

The LFP teachers in this study are products of this larger global and national context. To them, teaching was a respectable profession for women, enabling them to fulfill their domestic responsibilities. They perceived their education and knowledge of English giving them more status than the families of their first-generation students who came, as they put it, from "slum areas" and were "illiterate." Parents, for whom English was a pathway out of poverty, were acutely aware that their lack of "linguistic cultural capital" (Bourdieu & Passeron, 1990) disadvantaged them in two ways: They could neither challenge the status differential between them and the teachers nor support their child in their schoolwork other than by enrolling them in private tuition classes. By contrast, the relative parity in English fluency and social status between teachers and students combined with greater investment in infrastructure in the upper-tier schools contributed considerably to a higher quality of education for students in these schools.

English as a Factor in Maintaining Teacher-Centered Instruction

Despite the English-only policy in the LFP schools, teachers engaged in "guilty multilingualism" (Anderson & Lightfoot, 2018, p.16) and "teaching-in-translation" (Bhattacharya, 2013, p. 173), using a mixture of languages to teach their first-generation students who were learning English and learning in English simultaneously. However, the complexities of language, including the several languages and scripts at play, compelled teachers to "domesticate the pedagogy to suit local constraints" (Gupta, 2014, p. 3911) and utilize instructional strategies of rote learning they were familiar with and were the most efficacious within their contexts. To facilitate students' access to the content, they practiced choral recitation as they taught, and whittled down the content to "teachers' notes" as a written recitation script that students copied from the board.

Additionally, fiercely competitive educational environments from the demand for higher education outstripping supply have created a "marks race" (Majumdar & Mooij, 2012) with teachers teaching to the test to maximize students' chances of success on classroom assessments. Students who struggled to acquire literacy in the unfamiliar language or to maintain the pace of verbal and written recitation scripts were perceived by the teachers to be "slow learners," "back benchers," or "weak." By contrast, the upper-tier school students' familiarity with the language they were learning in facilitated, once again, access to a higher quality of education involving enriched level of content acquisition and critical thinking skills.

Supports for Academically Struggling Students

Large class sizes, rigid curriculum and minimal classroom resources beyond the textbook gave the teachers in the LFP schools little opportunity to provide individualized support. They offered free remedial classes where students were given time to copy notes they might have missed during the week. However, they relied primarily on the private tuition services outside of school to provide the supplemental support students needed and assumed parents would be willing to pay for these services (Ghosh & Bray, 2020). By contrast, the upper-tier schools offered in-school supports to academically struggling students. For an additional fee, students could avail of resource room or pull-out services through a trained special education teacher. Both teachers and parents preferred to use the term "dyslexia," if at all, because of its value-neutrality and the acknowledged stigma associated with learning disabilities.

There is a strong western-centric stance and emphasis on the medical model in the formal assessments used in the diagnostic clinics, both private and government (UNESCO, 2019; Velugu, no date, cited in UNESCO, 2019). The minimal, if any, adaptation of tests normed on western populations that were consistently used in the learning disability clinics privileged students fluent in English and familiar with American cultural references. Although some assessment measures have been developed for Indian populations, these are seldom used in diagnoses. Assessment that measure particular skills like visual-motor coordination, such as the Seguin Form Board, or visual perception, such as the Raven's Matrices Test, are perceived as being "culture-fair" because they are not language-based and used as proxies for IQ tests (Koshy et al., 2017).

India and the International Aid Agenda on Inclusive Education

The international aid agenda may have been well-intended to begin with. Certainly, the aim of universal education cannot be faulted. One might even argue that there is merit in the holy trinity of development—a free market economy, democratic government and human rights (Chang, 2003). However, there is an astonishing naiveté in (a) completely overlooking the role that colonialism played in capitalism that enabled the global North, today's developed countries, to get to where they are now at the expense of the global South, today's developing countries (Loomba, 2015); (b) assuming that the global North somehow has achieved the peak of developed perfection which all other countries would wish to emulate (Stiglitz, 2003); and (c) expecting that these goals would be met within a fraction of the time it took the global North (Chang, 2003).

Yet, it is within the context of these assumptions and expectations that the international aid agenda has compelled governments in the global South to establish educational systems that have structures and aims similar to those in the global North. The morphing of the role of a child from laborer in the field or factory to that of student, which occurred over centuries in the global North, led to the goal of universal education, Education For All by 2015, and more recently quality education for all by 2030, which is being asked of the global South within decades. And while the opportunities to access and inclusion of large populations of traditionally marginalized groups, such as the poor, girls, children with disabilities, and, in India, *Dalits*, in educational systems is undeniably an extraordinarily worthwhile target, it has also produced "Foucaultian 'docile bodies' that fit neatly into capitalist economies" and classroom spaces (Chocano, 2020, p. 9). Thus, we have schools that follow a calendar based on the rhythms of industrialized countries in countries which are still largely agrarian, with conflicting cycles of work and rest (Black, 2011). We have moved children from learning outdoors with their elders into buildings that students must sit inside, confined for several hours in the day (Ajmera, 2016; Dyer, 2001), and we blame students if they are distracted or unwilling to attend. We offer a curriculum that is expected to create a citizenry with national, and even global, values and skills for jobs that will lead children away from their natural communities in the belief that they are better off in the cities or overseas (Collin & Apple, 2010). We lure children into schools with the prospect that their lives will improve merely to introduce tests and other gate-keeping measures that allow only a privileged few to benefit (Boruah, 2017). We are ecstatic that enrollment rates are rising, assured that this must mean that parents believe in sending their children to school, when, in fact, parents have little choice but to ascribe to these parameters of schooling if they wish to aspire to economic and social mobility (Singal et al., 2019).

This dissonance of a "one-size-fits-all" mindset (Stiglitz, 2003) is particularly stark when we consider that the global North itself continues to struggle with its own imperfections in its educational systems. These very same structures have resulted in class-based inequities in access and opportunities to quality schooling in the US, which, given the strong correlation between class and race, have translated to low-income and non-white students being shut out of top tier universities, high paying, professional jobs, and other avenues of economic and social mobility (Collin & Apple, 2010). The situation has been further exacerbated by neoliberal economic reforms that have led away from pro-poor policies and toward public–private partnerships in education and other social sectors. As governments in the global South have implemented the structural adjustment advocated by the international agencies, these countries too have experienced wealth disparities, and a movement away from public investment in education. In India, private sector investment in education led to an increase in private schools in English to suit every pocket. The general perception that private schools are better than government also made them more attractive to parents. Although some have pointed to the proliferation of low-fee-paying schools as a desirable outcome of this agenda in that it affords access to an education to hitherto marginalized populations of children (Day et al., 2014; Tooley, 2009), the fact is these schools do not necessarily provide a quality education that ensures equitable economic and social benefits.

With English becoming a global *lingua franca*, and its increasing desirability as a required skill for employment in practically any capacity beyond a minimum-wage, unskilled job in India (Bhatia, 2017), the issue of medium of instruction adds another layer of complexity. Even the cashiers at a mid-scale coffee shop that I frequented spoke fluent English; considering all their clientele were Indian, this asset should not have been a requirement for their job. This expectation has forced poor parents, who might otherwise have sent their students to free, regional language, government-run schools, to pay for their child to attend a private school where they can acquire English. Private, low-fee-paying, English-medium schools exploit this vulnerability by offering immersion programs where students both learn English and learn in English simultaneously. They provide no supports for their poorly paid, and therefore equally poorly trained, teachers to teach English as a foreign language, which is veritably what it is for these students. Nor do they offer any assistance to students who might struggle academically under the enormity of this task, assuming they will avail of the shadow educational system of private tuition centers, or referring them to a clinic where they might end up being labeled learning disabled. There are no immediate benefits from this certificate, however. The document provides no educational recommendations that teachers could follow. On the contrary, the existence of this certificate only reifies the existence of "low IQ" for the teacher and validates the abrogation of their responsibility for teaching the child.

Following in the Footsteps

It is against this backdrop of universality of development that geodisability templates and global norming have contributed to the orthodoxy of inclusive education as the international aid agenda moved into the arena of disability and the education of children with disabilities in the1990s (Campbell, 2015; McDermott et al., 2011; Urwick & Elliott, 2010). Policies and legislation like the Salamanca Statement and the UN Convention on the Rights of People with Disabilities provided universal standards and norms for inclusive education. The categories of disability and the identification of who gets labeled, the types of services and the determination on where these services should be available, and the training of personnel and the decision on how schooling should occur that had emerged from the global North have become constructs of the global South (Croft, 2013). As a result, the patterns in the evolution of the area of learning disabilities are being transposed from the US and reproduced faithfully in India.

The similarities in the US and Indian context, such as wealth inequality and its strong correlation with linguistic diversity, have emphasized the inequities in school structures that privilege middle class and affluent students who are fluent in English. The competitiveness of the space race era that contributed to the social construction of the category of learning disabilities in the US, as more students began to be removed from general education, is akin to the competitive environment of globalization with its neoliberal market fundamentalist economic policies that is placing tremendous pressure on students in schools in India, resulting similarly in a group of students who are being marginalized from the educational mainstream. Since middle class and affluent students have more access to English on an everyday basis, it has been argued that English is not a confounding factor in their educational success and that students who do struggle academically are in fact learning disabled (Ramaa, 2000). However, the process by which these students in India are evaluated, assessed and labeled parallels the process in the US (Ahmad, 2015), despite the many problems inherent in it with regard to the validity of the instruments and the usefulness of the information gained from them. Additionally, almost none of these instruments has been normed to Indian populations, which raises similar and equally relevant questions as they continue to be used with students who are struggling academically (UNESCO, 2019). Further, unlike in the US, the fairly debilitating social stigma associated with the condition, notwithstanding the Bollywood movie, *Taare Zameen Par*, outweighs the benefits of any supports these students might receive.

Despite efforts to introduce the social model of disability that recognizes these categorizations as social constructs of a hostile and ableist environment, the overarching lens that prevails in India with regard to learning disability is that of the medical model. Acknowledgment of the social model extends only to a recognition of external factors, such as language, impinging on student failure, but does not deconstruct the systemic failure in placing a student in a situation where language becomes a barrier, rather than a facilitator, to learning. Instead, the fact that there are students who are struggling academically has been reified as a deficit intrinsic to the student and the assumption is that it is the student, not the educational structures, that needs fixing. Further, easy and immediate access to contemporary developments in the field in the global North, made possible by the Internet, has somewhat removed the development lag, which previously caused a delay in the transfer of ideas and practices. It has also facilitated much more vigorously the westward stance of NGOs and upper-tier schools in seeking solutions, to the point where they are spending their own revenues on bringing in international consultants and developing collaborations with agencies in the US, far beyond what the international aid technical assistance network could ever have imagined. The familiarity of professionals in the field in India with reading programs like Orton-Gillingham and the Sonday system, assessment measures like DIBELS and WRAT, and instructional outcomes like phonemic awareness and oral reading fluency, is strongly illustrative of how the field is progressing in India along the same lines as it has in the US, and an outcome of postcolonial globalization. Although upper-income children do benefit in the sense that they receive in-school supports and mandated modifications that help them to take school leaving exams as well as to qualify for colleges and universities abroad, the question is whether the label and the labeling process need to be so closely aligned with the western model, given the stigma attached to disability in India.

Education For All and the New Marginalization

In some ways, English becoming the global *lingua franca* is indicative of how truly global and transformative our current wave of globalization is from previous instances of globalization. Never before has the entire world been so enmeshed where a single language and culture has dominated interpersonal communication in practically all fields of commerce, science and technology, and the arts (Hansen, 2020; Parker, 2010). The richness of multilingualism in India and in many other former colonies around the world is rapidly being lost in a process of assimilation that began during colonization with the introduction of the imperial language as a form of maintaining political power. It continued, at the end of colonialism, with the creation of new nation states, where, in many, no one language stood out as a clear majority, and decisions about choosing a national language were fraught with the implications for power for the elite who spoke the colonial language. It has persisted in our contemporary times as movies, television, the Internet and social media have enabled American culture, the English language and even norms and values to overlay existing, traditional ways of life, the onslaught of which many national and local languages and cultures will probably never recover from (Breidlid, 2013; Sefa Dei & Jaimungal, 2018). In a country as multilingual as India, when did English become *primus inter pares*? Or, being a colonial legacy, did it always have the status of being first among equals with all the Indian languages?

The sweep of American empire occurred concurrently with efforts toward development and economic growth in the developing world and the international aid agenda, setting the standards for Education All. Although in 1953, UNESCO advocated for this education to be in the child's preferred language, the mother tongue, with a gradual introduction to the national language, it found, in 2003, that, due to rapidly accelerating globalization and profound political transformations, "in many countries that were previously under colonial regimes, the official language tends to be the language of the former colonizers" (UNESCO, 2003, p. 14). Similarly, although the Indian government has tried valiantly to protect its heritage of multilingualism, even as recently as 2020 in its National Education Policy, its movement toward neoliberal economic reform and market fundamentalism has undone much of its effort.

The business model that propelled the private sector to invest in low-fee-paying, English-medium schools was based on philanthropy or profit, not pedagogical principles. As a result, the schools that were established used the same immersion model that prevailed in the upper-tier schools, without any consideration of the fact that the students in the LFP schools came with little to no prior exposure to the English language. Research, on the other hand, shows that education in the child's preferred language, with bridged or gradual introduction to another language is more effective in helping students acquire proficiency in both languages (García, 2019; Skutnabb-Kangas, 2017). The teachers in these schools themselves were poorly trained and have little or no training in teaching English as a second or foreign language. Their own grasp of English was more colloquial and hybridized, what is often referred to as Indian English. They spent considerable instruction time explaining concepts in Hindi or Marathi, but in the ultimate analysis, they were helpless against the requirement that they teach students a pre-set curriculum entirely in English and always conscious of being berated for not complying with the "English only" policy. In response, they resorted to rote memorization as the most efficient—not effective—strategy for teaching. Students were expected to learn to read by recognizing words that were completely unfamiliar to them and not part of their everyday vocabulary. They were also expected to copy notes from the blackboard, learn them by heart, and essentially regurgitate them in the exam. So ubiquitous is this method that in the screening tool for learning disability adapted from the Ten Question Screening Inventory (TQSI), having difficulty being able to write notes from the blackboard was identified as one of the indicators for possible learning disability. Yet, concluding that the child may have a learning disability because they have difficulty with this task in such a pedagogically poor situation is tantamount to blaming the victim. Ironically, even as we seek to meet the goal of Education For all by including all previously marginalized children, the educational system is creating a new group of marginalized students.

There is something fundamentally flawed about an educational structure that creates more problems than resolves them, that breaks and blames children for a failure that is more structural than intrinsic to the child. The same arguments that have been made over decades in the US against educational processes that contributed to the marginalization of certain groups of students—and which have been largely unheeded—now prevail in the Indian situation. By making social and economic mobility contingent on acquiring this one arbitrary skill, whether it be literacy or fluency in English, the outcomes for low-income students in English-medium schools are much worse. In this immersion model of instruction, many students flounder as they try to acquire content in an entirely new language in a system that does not even consider that English is essentially a foreign language to them. Students who struggle to learn in this setting fare poorly because there are no in-school supports for them, and external supports, such as tuition centers, are equally focused on the content rather than the medium of instruction. For many of these students, the label of learning disability is the result of its conflation with language difference.

In some cases, the response to facilitate English language learning for "first generation English language learners" has been to introduce them to English at earlier and earlier stages of their education (Annamalai, 2004), despite evidence that this results in subtractive lingualism, where children lose their oral and, eventually reading, fluency in their preferred or first language, and become increasingly monolingual in English (García, 2019; Cummins, 1981). As a teacher educator described:

> We ran a series of programs, individual workshops at a (private, English-medium school) where many of the students are first generation learners and English speakers. We found that students in Standard 1 couldn't spell, read, write or speak in English. They (the school) wanted to catch them early, so they asked (us) to design a language program to teach English to their students at the pre-primary level, starting in upper KG. The program was very successful.... The medium of instruction was English. We found that the children were not comprehending until we made it English only. The feedback from the parents was amazing. (They) were so happy that their children were learning English so well... And the teachers were telling us that the upper KG who were going into standard 1 were ahead of the students who were already in Standard 1 (in English).

Annamalai (2004) notes that "the strategy of 'catching them young' with regard to learning English also finds political support" (p. 191) with several states changing their policy to start English teaching from the first grade instead of beginning it at post-primary level. Some of the parents in this study also mentioned sending their child to preschool so they would have learnt enough English to pass the Class 1 entrance exam to the LFP schools. One regretted not doing this with her older son because she "didn't know the system then," so he lost out and was now in a Marathi-medium school.

On the other hand, the immersion model in the LFP schools created a sense of guilt in the teachers when they attempted to bring in their students' preferred language into the classroom to facilitate their comprehension and learning (Anderson & Lightfoot, 2018; Bhattacharya, 2013), even though this is an appropriate strategy for rudimentary translanguaging (García, 2019). Further, the laborious process of translating the content usurps precious instructional time from the process of learning. Prior to the unit tests, a whole week was spent in a review of the curriculum. The students in the LFP schools were never exposed to the enriched learning that occurred in the upper-tier schools, and were never guaranteed the conditions even for basic learning.

Poor children do not receive a quality education, and the poorest of the poor are being excluded from the current system. Although they must receive an education by law, the education they do receive is not good enough to enable them to compete in the best higher education institutes where the medium of instruction is English (Dutt, 2019; Vulli, 2014). And although they may have also attended an English-medium school, the quality of this English—the accents, the fluency, the vocabulary—continues to set them apart as latecomers to the world of English speakers. It also renders them completely unfit to continue the traditional pursuits and lifestyles of their elders. The best of them, those who have passed the school leaving exam at Grade 10, might attend regional colleges and private institutes but are no nearer to achieving the kind of wealth and prestige they had aspired to (Bhatia, 2018). If they are lucky, they will get a job commensurate with their qualifications and make that giant leap from urban poor to lower middle class; but with the supply of labor vastly outstripping the demand for jobs, most do not. For others who have left the educational system earlier, the only option is vocational programs that might help them transition from being unskilled laborers in mines and construction to becoming semi-skilled laborers like gardeners and food preparers, and even greeters and bagboys in the hospitality industry, if they have English (Bhatia, 2018).

Despite the promise and accomplishments of Education For All, in the end, it has resulted in perpetuating the perennial elitist inequities of education for some. The social hierarchies continue to play out, privileging those who have always had access to education and linguistic cultural capital and benefitted from it, and disadvantaging those who have not (Bourdieu & Passeron, 1990). To what purpose, then, is the goal of quality Education For All if it results in a new marginalization?

The Two Indias, the New India

Contemporary globalization has changed India, arguably even more than colonization did. The race for coveted jobs, for wealth, for social status, has transmogrified into an extremely competitive educational system, where only the best, the crème de la crème can succeed (Majumdar & Mooij, 2012). Teachers spoke often of how the current fiercely competitive environment in schools was diminishing the cooperative, communal spirit of tolerance and accommodation that used to be a hallmark of Indian society, an attitude exemplified in the Kannada phrase "*sulpa* adjust *maadi*" ("adjust just a little") as an expected response to any inconvenient situation. Instead of this Asian collectivist, helping culture, there is now the very American individualistic belief of having to look out for oneself in a society in which peer pressure has begun to present itself in the form of bullying and teen suicides (Bhatia, 2018). As a counselor put it, "The Indian culture is just not tolerant anymore. The competiveness is eating at everybody; there's this sense of needing to create a niche for myself, I can't keep adjusting or I'll lose out." Administrators in upper-tier schools pointed to changes in parents' attitudes, "they are no longer willing to accept things as they are, but they now make demands on schools."

At the same time, globalization has also exacerbated age-old imbalances. In a society where hierarchies are so deeply embedded, it is difficult, if not impossible, to make arguments for inclusion on the basis of economics or on moral grounds. To argue that parallel structures are expensive or that students who don't make the grade should be accommodated instead of being pushed out is fairly meaningless. Parallel structures, however inefficient and expensive and unnecessary, have always existed, such as separate educational settings for Hindus, Christians, Muslims, Sikhs, and Jains, just as social inequities have, like inferior or non-existent schooling for Dalits (Chavan, 2013; Nambissan, 2012). As a school counselor said, "If we say, let's make it meaningful for all, the teachers respond with, 'why? There will always be some students who fail. Let the student go, there's 20 waiting to take their place!' We haven't been able to articulate an argument in favor of inclusion." As Ramaa (2000) points out, the multilingual context in India and the Three-Language Formula policy, which means that children are expected to learn to speak, read and write the regional language of the state, Hindi, and English, further complicates the scope of the issue.

The health crisis from the COVID-19 pandemic has only exacerbated this economic divide. Students who already had access to technology have benefitted from this, while students who had limited or no access have struggled to maintain their learning. Principals and other school administrators in various virtual forums attempting to present these challenges have described children borrowing a neighbor's phone to access WhatsApp, the social media platform their teacher is using to teach with. "Children are so eager to learn, and they are very resourceful," summed up one presenter. True, but is it fair to place such an inordinate responsibility for learning on a child? The concerns identified from the findings of this study suggest the need for some changes in the educational system.

Decolonization of Education and the Hegemony of English

"We see the challenge of decolonization as subverting the taken-for-granted, everyday assumptions deeply held by society that work to routinely reinforce White power and privilege," Sefa Dei and Jaimungal

(2018, p. 1) assert, adding “it is the assumed dominance, supremacy, and legitimacy of western knowledge that works to oppress, suppress – and delegitimize – other ways of knowing, thinking, being, living, and imagining… The hegemony of western science knowledge is revealed when we see Indigenous and colonized bodies dismissing (or outright refusing) knowledges and theorizations that are grounded in their local cultural contexts, histories and identities” (p. 3). They clarify that decolonization is not “simply a reaction or resistance to Europe” but that it “must delink colonial modes of knowledge production in order to decolonize knowledge and… education practices” (p. 3). Similarly, Grech (2011) recommends the need for eliciting local perceptions and understandings by adopting an assets or strengths-based lens. This involves identifying indigenous knowledge and embedded positive cultural values that offer a more inclusive response to difference.

In 1973, economist Ernst Schumacher, using India as an example, countered the movement toward globalization with the idea that “small is beautiful.” Capitalism, he pointed out, is essentially based on greed and makes people victims of their own aspirations, as embedded in the concept of “the pursuit of happiness,” where happiness and betterment are goals one aspires to but possibly never achieves. In his theory of economics “as if people mattered,” he proposed the possibility of an alternative to scale, a form of “Buddhist economics” that is smaller and therefore more sustainable, whereby development is village-based and defined by indigenous communities within the global South, not by countries and experts halfway across the world, geographically and economically. This does not mean that the “value-inequality” (Miles, 1985) built into the Indian social order is maintained, in that a cobbler’s son can aspire only to become the best cobbler and cannot aspire to any status other than a cobbler. According to Schumacher, this small scale would ensure against the vast inequities of wealth accumulation among a few individuals and the imbrication of wealth with social status, which have since become the hallmarks of globalization.

Since then other economists have also commented on the faulty assumptions of the international aid agenda on development and globalization (Chang, 2003; Stiglitz, 2003). Chang points to the absurdity of ignoring the role colonialism and slavery played in the development of capitalism in Europe and America in expecting the global South to achieve within a few decades what took the global North centuries to accomplish. Stiglitz (2007) writes about the fallacy of a "one-size-fits-all" model which overlooks the unique historical, political and social contexts of each country when international aid organizations attempt to impose templates for development. Expanding on these analyses, Chabbott (2015) asserts that the values which underlie the creation of the framework for Education For All are "consistent with Western Enlightenment norms, including rationality, individualism, professionalism, voluntarism, and global citizenship. This culture is neither the best nor likely the worst possible; rather, it is a reflection of the hegemony of Western science in commerce and in military affairs for at least 200 years" (p. 10).

Colonial modes of knowledge production and the hegemony of western science knowledge are evident in the stranglehold of English in language education, despite efforts by the Indian government to "demonetize" the effects of English within its Three-Language Formula policy in its most recent 2020 National Educational Policy. It is seen in the proliferation of English-medium private schools, the perception that acquisition of fluency in English is part of the aspirational lifestyle for middle and low-income families, and the brute fact that the English students might acquire in the LFP schools will continue to be perceived as being of poorer quality and therefore have less status than the English of the upper-tier schools.

The total immersion into an "English only" schooling environment that occurs in the English-medium LFP private schools is particularly deleterious for low-income students from linguistically diverse backgrounds for whom the language difference becomes conflated with a cognitive deficit, placing them at risk for being labeled learning disabled. The truth is these students are being immersed in academic content in a language they are almost completely unfamiliar with. The textbooks they worked with made no concessions to this fact: students in Class 3 were being introduced to words like *contamination*, *potable*, *interrogative* and *imperative*. The teachers' approach to teaching reading was through the Alphabet Spelling method which focuses on the letter rather than the letter sound, and emphasizes sight word recognition (Gupta, 2014; Sarma & Mukhtar, 2019). This is why, when shown the letter M, a student responded "M for *bandar* (monkey)." Another student, asked to read or sound out the word "bag," said "bucket-apple-giraffe." Even where a teacher was introducing students to phonics, the words (and pictures) associated with the letter sounds were equally unfamiliar to them, such as I for *igloo*, instead of, perhaps, the much more familiar word *India*, so that, once again, students had to resort to rote memory rather than picture association.

Teachers assumed students came ready to read in Class 1 and, concerned about the heaviness of the curriculum, focused entirely on teaching content. Once again, rote memorization was emphasized as teachers wrote the main points as notes on the blackboard after explaining the lesson in a mixture of English, Hindi and Marathi, and students were expected to copy these notes. This emphasis on recall and memorization was so embedded, assumed and ubiquitous in their teaching that it went unacknowledged. The few teachers who identified poor memory as a concern for potential learning difficulties in struggling students did so as a third or fourth behavior after "poor handwriting" or "slow in writing," "not paying attention," and "not bringing the correct notebooks."

Supports for struggling students consisted of "remedial classes" after school one or two days of the week, where they gave the students access to the class notes once again. Parents were not required to pay for this additional service, although they were often soundly berated by the teachers for not taking sufficient responsibility for their child's education during these sessions. However, most teachers relied on their students attending private tuition classes, a shadow education system that has sprung up parallel to the proliferation of private schools and the increase in competitiveness in school leaving exams (Ghosh & Bray, 2020). Not to be confused with remedial classes, these services were available, at a price, to all students, and have become *de rigueur*. Indeed, even academically successful students attended them for an extra boost in marks in the exams. At these classes or centers, the tutors, familiar with the curriculum, reviewed the lessons. Since the biggest value of this service was the small class size, the tutors were willing to re-teach a lesson if necessary with an individual child, but even they were reluctant to work with students who continued to struggle after a basic re-teach.

While some teachers in the LFP schools mentioned having recommended to parents in previous years that they move their child to a Marathi-medium school, where, they said, the students had begun to progress satisfactorily, only one acknowledged language as a factor causing students to struggle academically. All of them expressed chagrin about having to explain the content to their students in Hindi and/or Marathi, despite being an English-medium school, to facilitate their students' comprehension, even when I sought to reassure them that they were doing the right thing. And, even after these efforts, if "the students were not getting, what could they do?" After all, as they perceived it, these were students from the slum areas, whose parents were illiterate and unschooled about the demands of schooling.

In the documentary, *Schooling the World*, Carol Black (2011) identifies some of the consequences for children in rural areas of Ladakh, in the north-west of India, receiving this western form of education: loss of their own language, a disconnect with the elders of the community and their indigenous knowledges, physical and psychological separation from their heritage as they graduate and move into the cities to seek employment that is increasingly hard to get. The migrant families and the urban poor in this study have already experienced this disconnect from their roots in moving from the rural areas to the city. However, the introduction of English to the next generation causes further disruptions in relationships between parents and children, as parents are unable to help their children in any academic capacity in their learning, which can result in children coming to see themselves as superior to their parents (Bhatia, 2018).

Bilingual education strategies such as translanguaging, where the teachers provides opportunities for students to express themselves in their language of preference and the medium of instruction, have been recommended for students whose first language is not the medium of instruction (Garcia & Wei, 2014). Translanguaging involves more than direct translation of the content into the child's preferred language, the level of scaffolding the teachers in the LFP schools were providing, and moves beyond the binaries of language separation to leverage students' entire linguistic repertoires. Most significantly, it allows students to communicate and share their ideas, both among themselves and with the teachers, in their language of choice, whether it be English, Hindi, or Marathi, creating more flexible classroom spaces (Peden, 2018). Dual immersion bilingual education in primary school where students learn and acquire two languages, both conversational and academic, simultaneously with one language being the language of preference, also presents a viable option for the Indian context. Students would continue to be grounded in their mother tongue or language of preference while also acquiring fluency in either the regional language or English. This alternative could benefit both low-income and upper-income students.

Recognizing the Social Construction of Learning Disability

Colonial modes of knowledge production and the hegemony of western science knowledge are also evident in the construction of learning disabilities in India. It is embedded in the basic expectation that all children should be able to learn to read and write and do math, and, if they do not, they are inferior and this is somehow their fault (Kliewer & Biklen, 2007). The expectation itself arises from the international aid agenda of Education For All within the postcolonial framework that assumes a universality to what development means and that all Southern countries desire to be developed in the same way that the Northern countries are. And while an educated citizenry and universal literacy are laudable goals, the educational systems and structures for universal primary education that have been established toward this outcome also emerge from this western-centric framework. To some extent, considering there are several languages in the world that lack orthographies, there is a certain arbitrariness in the decision to make the cognitive skills of reading and writing a normative and universal goal. Would these skills have been such significant markers of competence if the radio, for instance, had been invented before the printing press?

In her analysis of social caste structures in racism, Wilkerson (2020) comments on the arbitrariness of these markers—height rather than color of skin could just as easily be the point of difference, because they are determined by those who possess the attribute of preference. In the supposed meritocracies of schools and our globalized world, the tyranny of these norms is expressed through the concept of intelligence, as measured through IQ tests and other equally arbitrary assessments (Parekh, 2017). The postcolonial echoes in the development of the field of learning disabilities in India with that of the US are clearly evident in the similarities of definition, with phrases such as exclusionary and inclusionary criteria and discrepancy factors, and the notion that the difficulty is intrinsic to the child (Ramaa, 2000; Unni, 2012). Initial investigations into the presence of reading and language difficulties in the latter half of the twentieth century focused on students in regional language-medium schools from average socio-economic backgrounds, and similarly identified difficulties in neuropsychological processes and information processing skills (Ramaa, 2000). Since most of the assessments at that time were conducted in the students' regional language, which we will presume was their preferred language, these difficulties are perceived as proof of innate deficits. This belief continues to persist within the field, as the numbers of children being identified as learning disabled continues to rise (Karanth, 2008; Ramaa, 2000; Thapa, 2008).

However, there is a circularity in the argument that "learning disabled students do poorly on these tests because they have learning disabilities" that often emerges in justification for this perception of difference, the labeling and the subsequent removal from general education of academically struggling students (Ramaa, 2000). It assumes a certain universality about the construct of IQ (Koshy et al., 2017; Serpell, 2000). As Kozleski (2020) asserts, "the process of special education identification and diagnosis bears little resemblance to medicine in which an illness is identified and treated. Instead, the process of determining who is referred for special education assessment and the outcomes of special education determination and categorization is mediated by social factors" (p. 345).

The interaction of language and learning when English is thrown into the mix, especially for low-income children for whom English is not an everyday language, accentuates the arbitrariness of this process. The same circular argument that many of these students are doing poorly in school because they have difficulty learning overlooks inequities in schooling, which include English-only school policies, poorly trained teachers and limited classroom resources, instructional strategies that focus on rote memorization and sight word reading, and large class sizes that preclude any individualized intervention. It also overlooks other social factors, such as limited to no access to English prior to and outside of school, governmental economic policies that compel rural populations to migrate to cities for employment where they are relegated to the urban fringes without basic amenities of housing, sanitation, water and electricity, international neoliberal economic policies that have built competitive, private schooling systems toward creating a global labor force, and the increasing emphasis on English as a necessary skill for any level of aspirational lifestyle resulting in the perception of English as a language of high status.

In conclusion, as Kozleski (2020) states, students "bring unique life experiences, cultural histories, languages, funds of knowledge, capacities, interests, and knowledge about how to 'do school.' Developmentally, culturally, experientially, students dynamically change because of their life circumstances AND the cultural experiences they have in school and elsewhere. Assessment processes identify learners who do not respond to the system as it is designed. Rather than change the system, the students are found lacking. There is nothing inclusive or emancipatory about this kind of education" (pp. 349–350). When we begin to see deficits in students, rather than the deficits in schooling systems, we are not being inclusive or emancipatory. When, rather than seeing the strengths of students who bring a wealth of languages to the classroom, we begin to see deficits because they have difficulty with the one language our era has deemed as the most prestigious, we are not being inclusive or emancipatory. When we create new perceptions of difference and systems for segregation and stigma and shame, in the extremely mistaken belief that we are helping students, once again, we are not being inclusive or emancipatory.

References

Ahmad, F. K. (2015). Exploring the invisible: Issues in identification and assessment of students with learning disabilities in India. *Transcience: A Journal of Global Studies, 6*(1), 91–107.

Ajmera, M. (2016). *Invisible children: Reimagining international development at the grassroots*. Palgrave Macmillan.

Anderson, J., & Lightfoot, A. (2018). Translingual practices in English classrooms in India: Current perceptions and future possibilities. *International Journal of Bilingual Education and Bilingualism*. https://doi.org/10.1080/13670050.2018.1548558

Annamalai, E. (2004). Medium of power: The question of English in education in India. In J. W. Tollefson & A. B. M. Tsui (Eds.), *Medium of instruction policies: Which agenda? Whose agenda?* (pp. 177–194). Routledge.

ASER. (2019). *Annual status of education report, 2018*. Pratham, Anthem Press.

Bangay, C., & Latham, M. (2013). Are we asking the right questions? Moving beyond the state vs non-state providers debate: Reflections and a case study from India. *International Journal of Educational Development, 33*, 244–252. https://doi.org/10.1016/j.ijedudev.2012.09.004

Bhatia, S. (2017, April 17). *Rub Out the Bias Behind Skin-Lightening Products: Sustained activism is needed to sway corporate exploitation of skin color perceptions*. US News and World Reports. https://www.usnews.com/news/best-countries/articles/2017-04-17/global-white-out-end-the-racism-that-reinforces-skin-lightening-products

Bhatia, S. (2018). *Decolonizing Psychology: Globalization, Social Justice, and Indian Youth Identities*. Oxford University Press.

Bhattacharya, U. (2013). Mediating inequalities: exploring English-medium instruction in a suburban Indian village school. *Current Issues in Language Planning, 14*(1), 164–184. https://doi.org/10.1080/14664208.2013.791236

Black, C. (Director). (2011). *Schooling the world: The white man's last burden* [Film]. Lost People Films. https://schoolingtheworld.org/film/

Boruah, P. (2017). Learning English in a low-cost semi-urban English-medium school in India: Challenges, interaction patterns and domains of use. In H. Coleman (Ed.), *Multilingualisms and Development* (pp. 289–306). British Council.

Bourdieu, P., & Passeron, J. C. (1990). *Reproduction in education, society and culture.* Sage.

Breidlid, A. (2013). *Education, indigenous knowledges and development in the global South: Contesting knowledges for a sustainable future.* Routledge.

Campbell, F. (2015). The terrain of disability law in Sri Lanka: Obstacles and possibilities for change. In S. Rao & M. Kalyanpur (Eds.), *South Asia and Disability Studies: Redefining boundaries and extending horizons* (pp. 73–98). Peter Lang.

Chabbott, C. (2015). *Institutionalizing Health and Education For All: Global goals, innovations and scaling up.* Teachers College Press.

Chang, H. J. (2003). *Kicking away the ladder: Development strategy in historical perspective.* Anthem Press.

Chattopadhay, T., & Roy, M. (2017). *Low-Fee Private Schools in India: The Emerging Fault Lines* (Working Paper 233). National Center for the Study of Privatization in Education. Teachers College, Columbia University. http://ncspe.tc.columbia.edu/center-news/low-fee-private-schools-in-india/

Chavan, D. (2013). *Language politics under colonialism: Caste, class and language pedagogy in Western India.* Cambridge Scholars Publishing.

Chocano, C. (2020, October 1). Distance learning, with shades of big brother. *The New York Times.* https://www.nytimes.com/2020/10/01/magazine/distance-learning-surveillance.html

Collin, R., & Apple, M. (2010). Globalization, schooling, and literacies in an unsettled age. In S. Tozer, B. P. Gallegos, A. Henry, M. Bushnell Greiner, & P. Groves Price (Eds.), *Handbook of research in the social foundations of education* (pp. 296–307). Routledge.

Croft, A. (2013). Promoting access to education for disabled children in low-income countries: Do we need to know how many disabled children there are? *International Journal of Educational Development, 33*, 233–243.

Cummins, J. (1981). Empirical and theoretical underpinnings of bilingual education. *Journal of Education.* https://doi.org/10.1177/002205748116300104

Day, L. C., McLoughlin, C., Aslam, M., Engel, J., Wales, J., Rawal, S., Batley, G., Kingdom, S., Nicolai, S., & Rose, P. (2014). *The role and impact of private school in developing countries: A rigorous review of the evidence.* Department for International Development.

Deb, S. (2011). *The beautiful and the damned: A portrait of the new India*. Faber.

Duhaney, L. M. G. (2014). Disproportionate representation in special education: A persistent stain on the field. In F. E. Obiakor & A. F. Rotatori (Eds.), *Contemporary perspectives in special education: Multicultural education for learners with special needs in the 21st century* (pp. 15–40). Information Age Publishing.

Dutt, Y. (2019). *Coming out as Dalit: A memoir*. Aleph.

Dyer, C. (2001). Nomads and Education for All: Education for development or domestication? *Comparative Education, 37*(3), 317–327. https://doi.org/10.1080/03050060120067802

García, O. (2019). Decolonizing foreign, second, heritage, and first languages: Implications for education. In D. Macedo (Ed.), *Decolonizing foreign language education: The misteaching of English and other colonial languages* (pp. 152–168). Routledge.

Garcia, O., & Wei, L. (2014). *Translanguaging: Language*. Palgrave Macmillan.

Ghosh, P., & Bray, M. (2020). School systems as breeding grounds for shadow education: Factors contributing to private supplementary tutoring in West Bengal, India. *European Journal of Education, 55*, 342–360. https://doi.org/10.1111/ejed.12412

Grech, S. (2011). Recolonising debates or perpetuated coloniality? Decentring the spaces of disability, development and community in the global south. *International Journal of Inclusive Education, 15*(1), 87–100.

Gupta, R. (2014). Change in teaching practices: Case of phonics instruction in India. *Procedia: Social and Behavioral Sciences, 116*, 3911–3915. https://doi.org/10.1016/j.sbspro.2014.01.865

Hansen, V. (2020). *The year 1000: When explorers connected the world—And globalization began*. Scribner.

Heyneman, S. P., & Stern, J. M. (2014). Low cost private schools for the poor: What public policy is appropriate? *International Journal of Educational Development, 35*, 3–15.

Jhingran, D. (2017). Education of equitable quality: All children reading. In M. J. Tiwary, S. Kumar, & A. K. Mitra (Eds.), *Dynamics of inclusive classrooms: Social diversity, inequality and school education in India* (pp. 81–104). Orient BlackSwan.

Juneja, N. (2017). Education in urban areas. In K. Kumar (Ed.), *Routledge handbook of education in India, debates, practices, and policies* (pp. 26–42). Routledge. https://www.routledgehandbooks.com/doi/10.4324/9781315107929.ch2

Karanth, P. (2008). Learning disability and language learning. In K. Thapa, G. M. van der Aalsvoort, & J. Pandey (Eds.), *Perspectives on learning disabilities in India: Current practices and prospects* (pp. 80–96). Sage.

Kliewer, C., & Biklen, D. (2007). Enacting literacy: Local understanding, significant disability, and a new frame for educational opportunity. *Teachers College Record, 109*(12), 2579–2600.

Koshy, B., Thomas, H. M. T, Samuel, P., Sarkar, R., Kendall, S., & Kang, G. (2017). Seguin Form Board as an intelligence tool for young children in an Indian urban slum. *Family Medicine and Community Health, 5*(4), 275–281. https://doi.org/10.15212/FMCH.2017.0118

Klingner, J. K., & Artiles, A. (2003). When should bilingual students be in special education? *Educational Leadership, 61*(2), 66–71.

Kozleski, E. B. (2020). Disrupting what passes as inclusive education: Predicating educational equity on schools designed for all. *The Educational Forum, 84*(4), 340–355. https://doi.org/10.1080/00131725.2020.1801047

Kumar, N. (2017). The technology and politics of teaching English. In M. J. Tiwary, S. Kumar, & A. K. Mitra (Eds.), *Dynamics of inclusive classrooms: Social diversity, inequality and school education in India* (pp. 183–202). Orient BlackSwan.

LaDousa, C. (2017). Mind the (language-medium) gap. In K. Kumar (Ed.), *Routledge handbook of education in India, debates, practices, and policies* (pp. 81–96). Routledge. https://www.routledgehandbooks.com/doi/10.4324/9781315107929.ch5.

Loomba, A. (2015). *Colonialism/Postcolonialism* (3rd ed.). Routledge.

Majumdar, M., & Mooij, J. (2012). The marks race: India's dominant education regime and new segmentation. In C. Sleeter, S. B. Upadhyay, A. K. Mishra, & S. Kumar (Eds.), *School education, pluralism and marginality: Comparative perspectives* (pp. 207–246). Orient Blackswan.

Manjrekar, N. (2013). Women school teachers in new times: Some preliminary reflections. *Indian Journal of Gender Studies, 20*(2), 335–356.

McDermott, R., Edgar, B., & Scarloss, B. (2011). Global norming. In A. Artiles, E. Kozleski, & F. Waitoller (Eds.), *Inclusive Education in five continents* (pp. 223–235). Harvard Education Press.

Miles, M. (1985). *Where there is no rehab plan: A critique of WHO's scheme for community based rehabilitation* [Monograph]. Hesperian Foundation.

Mohan, P. (2017). The road to English: The slow migration of the EWS child to elite India. In M. J. Tiwary, S. Kumar, & A. K. Mitra (Eds.), *Dynamics of inclusive classrooms: Social diversity, inequality and school education in India* (pp. 203–215). Orient BlackSwan.

Nambissan, G. B. (2012). Private schools for the poor: Business as usual? *Economic & Political Weekly, 47*(41), 51–58.

Panshikar, A. (2019). Specific learning disability: A hidden disability. In S. Chennat (Ed.), *Disability inclusion and inclusive education* (pp. 175–193). Springer. https://doi.org/10.1007/978-981-15-0524-9_9

Parekh, G. (2017). The tyranny of "ability". *Curriculum Inquiry, 47*(4), 337–343. https://doi.org/10.1080/03626784.2017.1383755

Parker, C. (2010). *Global interactions in the early modern age, 1400–1800*. University Press.

Peden, J. (2018). Decolonizing the geography classroom: A call to action for educators to reimagine pedagogy of place. In G. J. Sefa Dei & C. Jaimungal (Eds.), *Indigeneity and decolonial resistance: Alternatives to colonial thinking and practice* (pp. 77–94). Myers Press.

Ramaa, S. (2000). Two decades of research on learning disabilities in India. *Dyslexia, 6*, 268–283.

Sarangapani, P. M., & Winch, C. (2010). Tooley, Dixon and Gomathi on private education in Hyderabad: A reply. *Oxford Review of Education, 36*(4), 499–515. https://doi.org/10.1080/03054985.2010.495465

Sarma, P., & Mukhtar, M. (2019, November). *How to Educate One Billion Indians: The Akshar Education Model*. TEDx Talk. https://www.youtube.com/watch?v=IJdAb0mBYWE&feature=share&app=desktop

Sefa Dei, G. J., & Jaimungal, C. (2018). Indigeneity and decolonial resistance: An introduction. In G. J. Dei & C. Jaimungal (Eds.), *Indigeneity and decolonial resistance: Alternatives to colonial thinking and practice* (pp. 1–14). Myers Education Press.

Serpell, R. (2000). Intelligence and culture. In R. J. Sternberg (Ed.), *Handbook of intelligence* (pp. 549–577). Cambridge University Press. https://doi.org/10.1017/CBO9780511807947.026

Shenoy, S., & Wagner, R. K. (2019). Language and literacy practices that influence bilingual and bi-literate acquisition in L1 Kannada and L2 English in Bangalore, India. In R. M. Joshi & C. McBride (Eds.), *Handbook of literacy in Akshara orthography* (pp. 373–388). Springer. https://doi.org/10.1007/978-3-030-05977-4_19

Singal, N., Johansson, S. T., & Lynch, P. (2019). Education of children with disabilities: Changing landscape of new opportunities and challenges. In N. Singal, S. T. Johansson, & P. Lynch (Eds.), *Education and disability in the global South: New perspectives from Africa and Asia* (pp. 1–18). Bloomsbury Academic.

Sleeter, C. E. (1986). Learning disabilities: The social construction of a special education category. *Exceptional Children, 53*(1), 46–54.

Stiglitz, J. E. (2003). *Globalization and its discontents*. Norton.

Stiglitz, J. E. (2007). *Making globalization work: The next steps to social justice*. W.W. Norton.

Skutnabb-Kangas, T. (2017). Language rights and bilingual education. In O. García, A. M. Y. Lin & S. May (Eds.), *Bilingual and Multilingual Education* (3rd ed., pp. 51–64). Springer.

Thapa, K. (2008). Learning disabilities: Issues and concerns. In K. Thapa, G. M. van der Aalsvoort, & J. Pandey (Eds.), *Perspectives on learning disabilities in India: Current practices and prospects* (pp. 23–47). Sage.

Tooley, J. (2009). *The beautiful tree: A personal journey into how the world's poorest people are educating themselves*. Penguin.

UNESCO. (2003). *Education in a multilingual world*. Author.

UNESCO. (2019). *State of the education report for India 2019: Children with disabilities*. Author.

Unni, J.C. (2012). Specific learning disability and the amended "Persons with Disability Act". *Indian Pediatrics, 49*, 445–447. Retrieved from: http://medind.nic.in/ibv/t12/i6/ibvt12i6p445.pdf

Urwick, J., & Elliott, J. (2010). International orthodoxy versus national realities: Inclusive schooling and the education of children with disabilities in Lesotho. *Comparative Education, 46*(2), 137–150. https://doi.org/10.1080/03050061003775421

Vulli, D. (2014). English and medium of instruction: Dalit discourse in Indian education. *Research Journal of Educational Sciences, 2*(2), 1–6.

Wilkerson, I. (2020). *Caste: The origins of our discontents*. Random House.

Yengde, S. (2019). *Caste matters*. Penguin Viking.

References

10% of kids in India have learning disability: Experts say. (2012, January 27). *Times of India*. http://articles.timesofindia.indiatimes.com/2012-01-27/chennai/30669918_1_inclusive-education-disability-autism

40% of East Indians now more fluent in English than Marathi. (2018, July 8). *Times of India*, p. 4.

Adams, G. N., & Brown, S. M. (2007). *The six-minute solution: A reading fluency program*. Sopris West.

Addlakha, R. (Ed.). (2013). *Disability studies in India: Global discourses, local realities*. Routledge.

AfrAsia Bank. (2020). *Global Wealth Migration Review 2019*. https://e.issuu.com/embed.html?u=newworldwealth&d=gwmr_summary_presentation_2019

Ahmad, F. K. (2015). Exploring the invisible: Issues in identification and assessment of students with learning disabilities in India. *Transcience: A Journal of Global Studies, 6*(1), 91–107.

Ahsan, M. T., Deppeler, J. M., & Sharma, U. (2013). Predicting pre-service teachers' preparedness for inclusive education: Bangladeshi pre-service teachers' attitudes and perceived teaching-efficacy for inclusive education. *Cambridge Journal of Education, 43*(4), 517–535. https://doi.org/10.1080/0305764X.2013.834036

Air India Crash Investigators Focus on a Dangerous Runway and a Pilot's Actions. (2020, August 8). *The New York Times*. https://www.nytimes.com/2020/08/08/world/asia/india-plane-crash-dubai.html

M. Kalyanpur, *Development, Education and Learning Disability in India*, Palgrave Studies in Disability and International Development, https://doi.org/10.1007/978-3-030-83989-5

Ajmera, M. (2016). *Invisible children: Reimagining international development at the grassroots*. Palgrave Macmillan.

Alborz, S. R., & Miles, S. (2013). Establishing the foundations for an inclusive education system in Iraq reflection on findings from a nationwide survey. *International Journal of Inclusive Education, 17*(9), 965–987. https://doi.org/10.1080/13603116.2012.725776

Alessa, L. N. (2009). What is truth? Where science and traditional knowledge converge. In M. S. T. Williams (Ed.), *The Alaska native reader: History, culture, politics* (pp. 246–251). Duke University Press.

Almeida, R. (2020, June 10). *Tracing the roots of 25 English words with entirely Indian origins*. https://homegrown.co.in/article/44786/tracing-the-roots-of-25-english-words-with-entirely-indian-origins

Aloia, C. R., Gasevic, D., Yusuf, S., Teo, K., Chockalingam, A., Patro, B. K., Kumar, R., & Lear, S. A. (2013). Differences in perceptions and fast food eating behaviours between Indians living in high- and low-income neighbourhoods of Chandigarh India. *Nutrition Journal, 12*, 4. https://doi.org/10.1186/1475-2891-12-4

Alur, M. (2002). "They did not figure": Policy exclusion of disabled people in India. *International Journal of Inclusive Education, 6*, 101–112.

Anand, S. (2015). Corporeality and culture: Theorizing difference in the South Asian context. In S. Rao & M. Kalyanpur (Eds.), *South Asia and Disability Studies: Redefining boundaries and extending horizons* (pp. 158–174). Peter Lang.

Anderson, J. D. (1988). *The education of blacks in the south, 1860–1935*. The University of North Carolina Chapel Hill Press.

Anderson, J., & Lightfoot, A. (2018). Translingual practices in English classrooms in India: Current perceptions and future possibilities. *International Journal of Bilingual Education and Bilingualism*. https://doi.org/10.1080/13670050.2018.1548558

Annamma, S. A., Ferri, B. A., & Connor, D. J. (2018). Disability critical race theory: Exploring the intersectional lineage, emergence, and potential futures of discrit in education. *Review of Research in Education, 42*, 46–71. https://doi.org/10.3102/0091732X18759041

Annual State of Education Report (ASER). (2015). *Annual status of education report, 2014*. Pratham, Anthem Press.

Arne, E., & Ingstad, B. (Eds.). (2011). *Disability and poverty: A global challenge*. Policy Press.

Arvind, G. R. (2015). Institutional context, classroom discourse, and children's thinking: Pedagogy re-examined. In M. Thapan (Ed.), *Education and society: Themes, perspectives, practices* (pp. 283–308). Oxford University Press.

ASER. (2019). *Annual status of education report, 2018*. Pratham, Anthem Press.

Avramidis, E., & Norwich, B. (2002). Teachers' attitudes towards integration/inclusion: A review of the literature. *European Journal of Special Needs Education, 17*(2), 129–147.

Babu, M. J. P. (2018). Facilities provided by the primary education schools—A study in east Godavari district of Andhra Pradesh. *International Journal of Academic Research and Development, 3*(2), 1605–1607.

Baglieri, S., Bejoin, L. M., Broderick, A. A., Connor, D. J., & Valle, J. (2011). [Re]claiming "inclusive education" toward cohesion in educational reform: Disability studies unravels the myth of the normal child. *Teachers College Record, 113*(10), 2122–2154.

Baglieri, S., & Shapiro, A. (2017). *Disability Studies and the inclusive classroom: Critical practices for embracing diversity in education* (2nd ed.). Routledge.

Baine, D. (1988). *Handicapped children in developing countries: Assessment, curriculum, and instruction*. Vector Press.

Balachandran, K. (2014, August 21). *Turning the spotlight on the shadow teacher*. Deccan Herald. https://www.deccanherald.com/content/426628/turning-spotlight-shadow-teacher.html

Bandura, A. (1999). Social cognitive theory: An agentic perspective. *Asian Journal of Social Psychology, 2*(1), 21–41.

Bangay, C., & Latham, M. (2013). Are we asking the right questions? Moving beyond the state vs non-state providers debate: Reflections and a case study from India. *International Journal of Educational Development, 33*, 244–252. https://doi.org/10.1016/j.ijedudev.2012.09.004

Bellah, R. N., Madsen, R., Sullivan, W. M., Swidler, A., & Tipton, S. M. (2007). *Habits of the heart: Individualism and commitment in American life*. University of California Press.

Belliappa, J. L. (2014, October 15). *Gender inequities in school teaching: Reasons and repercussions*. Learning Curve. http://www.teachersofindia.org/en/article/gender-inequities-school-teaching-reasons-and-repurcussions

Béteille, A. (2012). *Caste and the citizen*. Oxford University Press.

Bhagat, R. B. (2017). *Migration and urban transition in India: Implications for development*. United Nations Secretariat. https://www.un.org/en/development/desa/population/events/pdf/expert/27/papers/V/paper-Bhagat-final.pdf

Bhatia, S. (2017, April 17). *Rub Out the Bias Behind Skin-Lightening Products: Sustained activism is needed to sway corporate exploitation of skin color perceptions*. US News and World Reports. https://www.usnews.com/news/best-countries/articles/2017-04-17/global-white-out-end-the-racism-that-reinforces-skin-lightening-products

Bhatia, S. (2018). *Decolonizing Psychology: Globalization, Social Justice, and Indian Youth Identities*. Oxford University Press.

Bhatnagar, N., & Das, A. K. (2014). Regular school teachers' concerns and perceived barriers to implement inclusive education in New Delhi India. *International Journal of Instruction, 7*(2), 89–102.

Bianchi, C. (2018). Reflections on the implications of Western theories on indigenous populations: Decolonizing and indigenizing the classroom. In G. J. Dei & C. Jaimungal (Eds.), *Indigeneity and decolonial resistance: Alternatives to colonial thinking and practice* (pp. 77–94). Myers Education Press.

Black, C. (Director). (2011). *Schooling the world: The white man's last burden* [Film]. Lost People Films. https://schoolingtheworld.org/film/

Blanchett, W. J., Klingner, J. K., & Harry, B. (2009). The intersection of race, culture, language, and disability: Implications for urban education. *Urban Education, 44*(4), 389–409. https://doi.org/10.1177/0042085909338686

Bogdan, R., & Biklen, S. K. (2006). *Qualitative research for education: An introduction to theories and methods* (5th ed.). Pearson.

Boo, K. (2012). *Behind the beautiful forevers: Life, death and hope in a Mumbai undercity*. Random House.

Boroditsky, L. (2017). *How language shapes the way we think.* TED Talk https://www.ted.com/talks/lera_boroditsky_how_language_shapes_the_way_we_think?language=en

Boruah, P. (2017). Learning English in a low-cost semi-urban English-medium school in India: Challenges, interaction patterns and domains of use. In H. Coleman (Ed.), *Multilingualisms and Development* (pp. 289–306). British Council.

Bourdieu, P., & Passeron, J. C. (1990). *Reproduction in education, society and culture*. Sage.

Bowen, G. A. (2009). Document analysis as a qualitative research method. *Qualitative Research Journal, 9*(2), 27–40.

Bowles, S., & Gintis, H. (2011). *Schooling In capitalist America: Educational reform and the contradictions of economic life*. Haymarket Books.

Breidlid, A. (2013). *Education, indigenous knowledges and development in the global South: Contesting knowledges for a sustainable future*. Routledge.

Brissett, N. O. M., & Bailey, C. Y. (2021). Small Axe, episode 5, "Education" directed by Steve McQueen, produced by Anita Overland and Mike Elliott, and edited by Chris Dickens and Steve McQueen. *Comparative Education Review, 65*(2), 380–381.

By 2027, India will triple its billionaire count. (2018, July 12). *Times of India*, p. 7.

Campbell, F. (2015). The terrain of disability law in Sri Lanka: Obstacles and possibilities for change. In S. Rao & M. Kalyanpur (Eds.), *South Asia and Disability Studies: Redefining boundaries and extending horizons* (pp. 73–98). Peter Lang.

Canagarajah, S. (2006). Changing communicative needs, revised assessment objectives: Testing English as an international language. *Language Assessment Quarterly, 3*(3), 229–242.

Celarent, B. (2011). Caste and Race in India by G. S. Ghurye. *American Journal of Sociology, 116*(5), 1713–1719.

Center for Equity Studies (CES). (2014). *India exclusion report, 2013–14*. Books for Change.

Chabbott, C. (2015). *Institutionalizing Health and Education For All: Global goals, innovations and scaling up*. Teachers College Press.

Chakravarty, M. (4 June, 2020). *Richest 10% of Indians own over 3/4th of wealth in India*. https://www.livemint.com/Money/iH2aBEUDpG06hM78diSSEJ/Richest-10-of-Indians-own-over-34th-of-wealth-in-India.html

Chang, H., & Grabel, I. (2004). *Reclaiming development: An alternative economic policy manual*. Palgrave Macmillan.

Chang, H. J. (2003). *Kicking away the ladder: Development strategy in historical perspective*. Anthem Press.

Charmaz, K. (2014). *Constructing grounded theory* (2nd ed.). Sage.

Chaterjee, P. (1991). Whose imagined community? *Millennium: Journal of International Studies, 20*(3), 521–525. https://doi.org/10.1177/03058298910200030601

Chaterjee, P. (1993). *The nation and its fragments: Colonial and postcolonial histories*. Princeton University Press.

Chattopadhay, T., & Roy, M. (2017). *Low-Fee Private Schools in India: The Emerging Fault Lines* (Working Paper 233). National Center for the Study of Privatization in Education. Teachers College, Columbia University. http://ncspe.tc.columbia.edu/center-news/low-fee-private-schools-in-india/

Chavan, D. (2013). *Language politics under colonialism: Caste, class and language pedagogy in Western India*. Cambridge Scholars Publishing.

Chennat, S. (2019a). Disability studies: The context. In S. Chennat (Ed.), *Disability inclusion and inclusive education* (pp. 1–19). Springer. https://doi.org/10.1007/978-981-15-0524-9_1

Chennat, S. (2019b). Sociology of disability. In S. Chennat (Ed.), *Disability inclusion and inclusive education* (pp. 21–37). Springer. https://doi.org/10.1007/978-981-15-0524-9_2

Chennat, S. (2019c). Conceptualizing disability inclusion. In S. Chennat (Ed.), *Disability inclusion and inclusive education* (pp. 39–62). Springer. https://doi.org/10.1007/978-981-15-0524-9_3

Chocano, C. (2020, October 1). Distance learning, with shades of big brother. *The New York Times*. https://www.nytimes.com/2020/10/01/magazine/distance-learning-surveillance.html

Coleman, H. (2017). Development and multilingualism: An introduction. In H. Coleman (Ed.), *Multilingualisms and development: Selected proceedings of the 11th language and development conference* (pp. 15–34). British Council. http://www.langdevconferences.org/publications/2015-NewDelhi/MultilingualismsandDevelopment-Coleman-ed-completepublications.pdf

Collin, R., & Apple, M. (2010). Globalization, schooling, and literacies in an unsettled age. In S. Tozer, B. P. Gallegos, A. Henry, M. Bushnell Greiner, & P. Groves Price (Eds.), *Handbook of research in the social foundations of education* (pp. 296–307). Routledge.

Connor, D. J., Valle, J. W., & Hale, C. (2015). *Practicing disability studies in education: Acting towards social change*. Peter Lang.

Corbin, J., & Strauss, A. (2015). *Basics of qualitative research: Techniques and procedures for developing grounded theory* (4th ed.). Sage.

Creswell, J. W. (2007). *Qualitative inquiry and research design: Choosing among five traditions*. Sage.

Croft, A. (2013). Promoting access to education for disabled children in low-income countries: Do we need to know how many disabled children there are? *International Journal of Educational Development, 33*, 233–243.

Cummins, J. (1981). Empirical and theoretical underpinnings of bilingual education. *Journal of Education*. https://doi.org/10.1177/002205748116300104

Dalrymple, W. (June 29, 2015). The great divide: The violent legacy of Indian Partition. *The New Yorker*. https://www.newyorker.com/magazine/2015/06/29/the-great-divide-books-dalrymple

Das, J. P. (2008). Foreword. In K. Thapa & G. M. van der Aalsvoort (Eds.), *Perspectives on learning disabilities in India: Current practices and prospects* (pp. ii–v). Sage.

Datt, G., & Wang, L. C. (2020). Schooling and learning deficits: A simple unified measurement framework. *Comparative Education Review, 64*(3). http://doi.org/0010-4086/2020/6403-0008

Day, L. C., McLoughlin, C., Aslam, M., Engel, J., Wales, J., Rawal, S., Batley, G., Kingdom, S., Nicolai, S., & Rose, P. (2014). *The role and impact of private school in developing countries: A rigorous review of the evidence*. Department for International Development.

de Araujo, Z. (2017). Connections between secondary mathematics teachers' beliefs and their selection of tasks for English language learners. *Curriculum Inquiry, 47*(4), 363–389. https://doi.org/10.1080/03626784.2017.1368351

Deb, S. (2011). *The beautiful and the damned: A portrait of the new India*. Faber.

Desai, S., & Thorat, A. (2013). Social inequalities in education. In I. D. F. C. Foundation (Ed.), *India Infrastructure Report 2012: Private Sector in Education* (pp. 44–50). Routledge.

Dhanu, G., Havale, R., Shrutha, S. P., Quazi, N., Shafna, T. P., & Ahmed, A. (2019). Assessment of Intelligence Quotient using Raven's Coloured Progressive Matrices among school children of Hyderabad Karnataka region and its correlation with prevalence of dental caries. *Journal of the Indian Society of Pedodontics and Preventive Dentistry, 37*(1), 25–30. https://doi.org/10.4103/JISPPD.JISPPD_236_18

Dudley-Marling, C. (2004). The social construction of learning disabilities. *Journal of Learning Disabilities, 37*(6), 482–489.

Duhaney, L. M. G. (2014). Disproportionate representation in special education: A persistent stain on the field. In F. E. Obiakor & A. F. Rotatori (Eds.), *Contemporary perspectives in special education: Multicultural education for learners with special needs in the 21st century* (pp. 15–40). Information Age Publishing.

Dyer, C. (2001). Nomads and Education for All: Education for development or domestication? *Comparative Education, 37*(3), 317–327. https://doi.org/10.1080/03050060120067802

Dyslexia Association of India. (2011). *Home page*. http://www.dyslexiaindia.org.in/

DysTEFL2. (2021). *Dyslexia for Teachers of English as a Foreign Language 2*. http://dystefl2.uni.lodz.pl/

Easterly, W. (2006). *The white man's burden: Why the west's efforts to aid the rest have done so much ill and so little good*. Penguin.

ELT well. (2021). *Inclusive language teaching*. http://eltwell.com/resources/inclusive-language-teaching/

EPW Engage. (2021, April 23). *English language education in India: How aspirations for social mobility shape pedagogy*. https://www.epw.in/engage/article/english-language-education-india-aspirations-pedagogy

Erevelles, N. (2011). *Disability and difference in global contexts: Enabling a transformative body politic*. Palgrave Macmillan.

Fannon, F. (2004). *The wretched of the earth*. Grove Press.

Ferri, B. A., & Connor, D. J. (2006). Tools of exclusion: Race, disability and (re)segregated education. *Teachers College Record, 107*(3), 453–474.

Florian, L., Hollenweger, J., Simeonsson, R. J., Wedell, K., Riddell, S., Terzi, L., & Holland, A. (2006). Cross-cultural perspectives on the classification of children with disabilities: Part I. Issues in the classification of children with disabilities. *The Journal of Special Education, 40*(1), 36–45.

Foucault, M. (1972). *The archaeology of knowledge and the discourse on language*. Pantheon Books.

Frank, M. C., Everett, D. L., Fedorenko, E., & Gibson, E. (2008). Number as a cognitive technology: Evidence from Pirahã language and cognition. *Cognition, 108*, 819–824. https://doi.org/10.1016/j.cognition.2008.04.007

Frey, N., & Fisher, D. (2009). The release of successful instruction. *Principal Leadership, 9*(6), 18–22.
Friedman, T. L. (2005). *The world is flat: A brief history of the twenty-first century.* Farrar, Straus & Giroux.
Gabel, S. (2004). South Asian Indian cultural orientations towards mental retardation. *Mental Retardation, 42*(1), 12–25.
Ganapathy-Coleman, H. (2014). 'Teachers are always good. Children have flaws': Memories of school and learning in the narratives of low-income parents in India. *Psychology and Developing Societies, 26*(1), 29–58.
Gandhe, S. K. (2004). *External evaluation of Janshala programme: synthesised report* (Pune, Indian Institute of Education).
García, O. (2019). Decolonizing foreign, second, heritage, and first languages: Implications for education. In D. Macedo (Ed.), *Decolonizing foreign language education: The misteaching of English and other colonial languages* (pp. 152–168). Routledge.
García, O., Johnson, S., & Seltzer, K. (2017). *The translanguaging classroom: Leveraging student bilingualism for learning.* Caslon.
Garcia, O., & Wei, L. (2014). *Translanguaging: Language.* Palgrave Macmillan.
Garrett, L. (2000). Filth and decay: Pneumonic plague hits India and the world ill responds. In L. Garrett (Ed.), *Betrayal of trust: The collapse of global public health* (pp. 15–48). Hyperion.
Ghai, A. (2002). Disability in the Indian context: Post-colonial perspectives. In M. Corker & T. Shakespeare (Eds.), *Disability/Postmodernity: Embodying disability theory* (pp. 88–100). Continuum.
Ghai, A. (2020). Ignorance of disability: Some epistemological questions. In N. Mehrotra (Ed.), *Disability Studies in India* (pp. 75–91). Springer. https://doi.org/10.1007/978-981-15-2616-9_1
Ghurye, G. S. (1932). *Caste and race in India.* Kegan Paul, Trench, Trübner.
Giroux, H. A. (1983). *Theory and resistance in education: A pedagogy for the opposition.* Bergin.
Giroux, H. A. (2002). Democracy, freedom, and justice after September 11th: Rethinking the role of educators and the politics of schooling. *Teachers College Record, 104*, 1138–1162.
Glesne, C. (2016). *Becoming qualitative researchers: An introduction* (5th ed.). Pearson.
A 'Goddess of English' for India's down-trodden. (2011, February 15). BBC News. http://www.bbc.com/news/world-south-asia-12355740
Goodley, D. (2011a). Introduction: Global disability studies. In D. Goodley (Ed.), *Disability studies: An interdisciplinary introduction* (pp. 1–21). Sage.
Goodley, D. (2011b). Education: Inclusive disability studies. In D. Goodley (Ed.), *Disability studies: An interdisciplinary introduction* (pp. 139–156). Sage.

Gould, S. J. (1996). *The mismeasure of man*. W.W. Norton.

Grech, S. (2009). Disability, poverty and development: Critical reflections on the majority world debate. *Disability & Society, 24*, 771–784.

Grech, S. (2011). Recolonising debates or perpetuated coloniality? Decentring the spaces of disability, development and community in the global south. *International Journal of Inclusive Education, 15*(1), 87–100.

Grosfoguel, R. (2013). The structure of knowledge in westernized universities epistemic racism/sexism and the four genocides/epistemicides of the long 16th Century. *Human Architecture: Journal of the Sociology of Self-Knowledge, 11*(1), 73–90.

Guha, R. (2020, June 9). *Traits of Modi that have cost India dearly*. NDTV. https://www.ndtv.com/opinion/3-traits-of-modi-that-have-cost-india-dearly-by-ramachandra-guha-2242669

Gupta, R. (2014). Change in teaching practices: Case of phonics instruction in India. *Procedia: Social and Behavioral Sciences, 116*, 3911–3915. https://doi.org/10.1016/j.sbspro.2014.01.865

Gurumurthy, A. A. (2021, May 28). What I learned about caste while discussing the PSBB incident with family, friends. *The Wire*. https://livewire.thewire.in/campus/psbb-incident-casteism-oppression/

Gusain, L. (2012). The effectiveness of establishing Hindi as a national language. *Georgetown Journal of International Affairs, 13*(1), 43–50.

Habib, D. (2017). *Intelligent lives*. New Hampshire Charitable Foundation. Trailer. https://www.youtube.com/watch?v=jjyENb2qbu0

Hallberg, L. R.-M. (2006). The "core category" of grounded theory: Making constant comparisons. *International Journal of Qualitative Studies on Health and Well-Being, 1*(3), 141–148. https://doi.org/10.1080/17482620600858399

Hammad, T., & Singal, N. (2015). Disability, gender, and education: Exploring the impact of education on the lives of women with disabilities in Pakistan. In S. Rao & M. Kalyanpur (Eds.), *South Asia and Disability Studies: Redefining boundaries and extending horizons* (pp. 197–228). Peter Lang.

Hansen, V. (2020). *The year 1000: When explorers connected the world—And globalization began*. Scribner.

Harber, C. (2014). *Education and International Development: Theory, practice and issues*. Symposium Books.

Hardt, M., & Negri, A. (2000). *Empire*. Harvard University Press.

Härmä, J. (2011). Low cost private schooling in India: Is it pro poor and equitable? *International Journal of Educational Development, 31*, 350–356. https://doi.org/10.1016/j.ijedudev.2011.01.003

Harry, B. (2014). The disproportionate placement of ethnic minorities in special education. In L. Florian (Ed.), *The SAGE Handbook of special education* (Vol. 1, 2nd ed., pp. 73–95). Sage.

Harry, B. (2020). *Childhood disability, advocacy, and inclusion in the Caribbean: A Trinidad and Tobago case study*. Springer.

Hettiarachchi, S., & Das, A. (2014). Perceptions of 'inclusion' and perceived preparedness among school teachers in Sri Lanka. *Teaching and Teacher Education, 43*, 143–153. https://doi.org/10.1016/j.tate.2014.07.003

Heyneman, S. P., & Stern, J. M. (2014). Low cost private schools for the poor: What public policy is appropriate? *International Journal of Educational Development, 35*, 3–15.

Hilowle, S. (2018). Education in Somalia: The role of international organizations in formal education. In G. J. Dei & C. Jaimungal (Eds.), *Indigeneity and decolonial resistance: Alternatives to colonial thinking and practice* (pp. 139–152). Myers Education Press.

Hindi mother tongue of 44% of Indians, Marathi 3rd most spoken. (2018, June 27). *Times of India*, p. 11.

Illich, I. (1971). *Deschooling society*. Marion Boyars.

Indian campaign of Alexander the Great. (2020, May 6). *Wikipedia*. https://en.wikipedia.org/wiki/Indian_campaign_of_Alexander_the_Great

Indian Railways jobs: 2.4 crore candidates apply for 1.2 lakh vacancies. (2019, November 7). *Business Today*. https://www.businesstoday.in/sectors/jobs/indian-railways-jobs-2-4-crore-candidates-apply-1-2-lakh-vacancies/story/389097.html

Individuals with Disabilities Education Act (IDEA). (2014). (Sec. 300.8 (c) (10). https://sites.ed.gov/idea/regs/b/a/300.8/c/10

In Puri, priests block President Kovind's way, shove First Lady. (2018, June 27). *Times of India*, p. 9.

International Labor Organization. (2002). *Disability and poverty reduction strategies: How to ensure that access of persons with disabilities to decent and productive work is part of the poverty reduction strategy process* (Discussion Paper). Author.

Jackson, R. F., & Lee, C. M. (1992). *Faking it: A look into the mind of a creative learner*. Heinemann.

Jallade, L., Radi, M., & Cuenin, S. (2001). *National education policies and programmes and international cooperation: What role for UNESCO?* UNESCO.

Jangira, N. K. (2002). Special educational needs of children and young adults: An unfinished agenda. In M. Alur & S. Hegarty (Eds.), *Education and children with special needs: From segregation to inclusion* (pp. 67–76). Sage.

Jelas, Z. M., & Ali, M. M. (2014). Inclusive education in Malaysia: Policy and practice. *International Journal of Inclusive Education, 18*(10), 991–1003. https://doi.org/10.1080/13603116.2012.693398

Jha, M. M. (2004). Inclusive education and the common school in India. In C. S. Mohapatra (Ed.), *Disability management in India: Challenges and commitments* (pp. 160–171). National Institute for the Mentally Handicapped.

Jhingran, D. (2017). Education of equitable quality: All children reading. In M. J. Tiwary, S. Kumar, & A. K. Mitra (Eds.), *Dynamics of inclusive classrooms: Social diversity, inequality and school education in India* (pp. 81–104). Orient BlackSwan.

Joseph, T. (2018, April 3). How we, the Indians, came to be. *The Quint*. https://www.thequint.com/voices/opinion/genomic-study-vedic-aryan-migration-dravidian-languages-sanskrit

Joseph, T. (2019, September 13). There's no confusion: The new reports clearly confirm 'Arya' migration into India. *The Hindu*. https://www.thehindu.com/society/history-and-culture/theres-no-confusion-the-new-reports-clearly-confirm-arya-migration-into-india/article29409611.ece

Juneja, N. (2017). Education in urban areas. In K. Kumar (Ed.), *Routledge handbook of education in India, debates, practices, and policies* (pp. 26–42). Routledge. https://www.routledgehandbooks.com/doi/10.4324/9781315107929.ch2

Kalia, V., & Reese, E. (2009). Relations between Indian children's home literacy environment and their English oral language and literacy skills. *Scientific Studies of Reading, 13*(2), 122–145. https://doi.org/10.1080/10888430902769517

Kalyanpur, M. (1996). The influence of western special education on community-based services in India. *Disability & Society, 11*(2), 249–270.

Kalyanpur, M. (2008a). The paradox of majority under-representation in special education in India: Constructions of difference in a developing country. *Journal of Special Education, 42*(1), 55–64.

Kalyanpur, M. (2008b). Equality, quality and quantity: Challenges in inclusive education policy and service provision in India. *International Journal of Inclusive Education, 12*(3), 243–263.

Kalyanpur, M. (2014). Distortions and dichotomies in inclusive education for children with disabilities in Cambodia in the context of globalization and international development. *International Journal of Disability, Development and Education, 61*(1), 80–94, 15p. https://doi.org/10.1080/1034912X.2014.878546

Kalyanpur, M. (2015). Mind the gap: The evolution of special education policy and practice in India in the context of globalization. In S. Rao & M. Kalyanpur (Eds.), *South Asia and Disability Studies: Redefining boundaries and extending horizons* (pp. 49–72). Peter Lang.

Kalyanpur, M. (2020). Inclusive and special education in Asia. In U. Sharma & S. Salend (Eds.), *Oxford Research Encyclopedia, Education*. Oxford University

Press. https://doi.org/10.1093/acrefore/9780190264093.013.ORE_EDU-01262.R1

Kalyanpur, M., & Gowramma, I. P. (2007). Cultural barriers to South Indian families' access to services and educational goals for their children with disabilities. *Journal of the International Association of Special Education, 8*(1), 69–82.

Kalyanpur, M., & Harry, B. (2012). *Cultural reciprocity in special education: Building reciprocal family-professional relationships*. Brookes.

Karande, S., Sholapurwala, R., & Kulkarni, M. (2011). Managing specific learning disability in schools in India. *Indian Pediatrics, 48*, 517–520.

Karanth, P. (2008). Learning disability and language learning. In K. Thapa, G. M. van der Aalsvoort, & J. Pandey (Eds.), *Perspectives on learning disabilities in India: Current practices and prospects* (pp. 80–96). Sage.

Karna, G. N. (2001). *Disability studies in India: Retrospects and prospects*. Gyan Publishing.

Keenan, J. M., & Betjemann, R. S. (2006). Comprehending the Gray Oral Reading Test without reading it: Why comprehension tests should not include passage-independent items. *Scientific Studies of Reading, 10*(4), 363–380. https://doi.org/10.1207/s1532799xssr1004_2

Kelly, G. P., & Altbach, P. G. (1984). The four faces of colonialism. In G. P. Kelly & P. G. Altbach (Eds.), *Education and the Colonial Experience* (pp. 1–5). Transaction.

Khubchandani, L. M. (1975). Dilemmas of language transition: Challenges to language planning in India. *ERIC Document Reproduction Service No. ED, 116*, 474.

Klees, S. J. (2017, August 3). *Financing Universal Education*. https://www.project-syndicate.org/commentary/financing-universal-education-by-steven-j--klees-2017-08

Kliewer, C., & Biklen, D. (2007). Enacting literacy: Local understanding, significant disability, and a new frame for educational opportunity. *Teachers College Record, 109*(12), 2579–2600.

Kliewer, C., & Fitzgerald, L. M. (2001). Disability, schooling, and the artifacts of colonialism. *Teachers College Record, 103*(3), 450–470.

Klingner, J. K., & Artiles, A. (2003). When should bilingual students be in special education? *Educational Leadership, 61*(2), 66–71.

Kolb, S. M. (2012). Grounded theory and the constant comparative method: Valid research strategies for educators. *Journal of Emerging Trends in Educational Research and Policy Studies (JETERAPS), 3*(1), 83–86.

Kopf, D. (1984). Orientalism and the Indian educated elite. In G. P. Kelly & P. G. Altbach (Eds.), *Education and the Colonial Experience* (pp. 117–136). Transaction Books.

Kormos, J., & Nijakowska, J. (2017). Inclusive practices in teaching students with dyslexia: Second language teachers' concerns, attitudes and self-efficacy beliefs on a massive open online learning course. *Teaching and Teacher Education, 68*, 30–41. https://doi.org/10.1016/j.tate.2017.08.005

Kozleski, E. B. (2020). Disrupting what passes as inclusive education: Predicating educational equity on schools designed for all. *The Educational Forum, 84*(4), 340–355. https://doi.org/10.1080/00131725.2020.1801047

Kozleski, E. B., Artiles, A. J., & Waitoller, F. R. (2015). Introduction: Equity in inclusive education. In A. J. Artiles, E. B. Kozleski, & F. R. Waitoller (Eds.), *Inclusive education: Examining equity on five continents* (pp. 1–14). Harvard Education Press.

Krishnamurthy, V. (2019, November). *How we can create a more inclusive world for kids with disabilities*. TED Talks India: Nayi Baat. https://www.ted.com/talks/vibha_krishnamurthy_how_we_can_create_a_more_inclusive_world_for_kids_with_disabilities#t-799815

Kumar, N. (2017). The technology and politics of teaching English. In M. J. Tiwary, S. Kumar, & A. K. Mitra (Eds.), *Dynamics of inclusive classrooms: Social diversity, inequality and school education in India* (pp. 183–202). Orient BlackSwan.

LaDousa, C. (2017). Mind the (language-medium) gap. In K. Kumar (Ed.), *Routledge handbook of education in India, debates, practices, and policies* (pp. 81–96). Routledge. https://www.routledgehandbooks.com/doi/10.4324/9781315107929.ch5

Ladwig, J. G., & McPherson, A. (2017). The anatomy of ability. *Curriculum Inquiry, 47*(4), 344–362. https://doi.org/10.1080/03626784.2017.1368352

Lawrence, E. N. (2017, May 23). *Let's Talk About Racism | 'I am tired of being a black man in India'*. https://www.hindustantimes.com/india-news/let-s-talk-about-racism-kallu-was-the-first-word-this-african-man-learnt-in-india/story-f04HSkREMwyIBh3GlO3DDM.html

Leach, D., & Helf, S. (2016). Revisiting the regular education initiative: Multi-tiered systems of support can strengthen the connection between general and special education. *Journal of the American Academy of Special Education Professionals*, 116–124.

Le Fanu, G. (2013). Reconceptualising inclusive education in international development. In L. Tikly & A. Barrett A (Eds.), *Education, quality and social justice in the global south* (pp. 40–55). Routledge.

Leiberman, R. H. (Director). (2017). *Angkor awakens: A portrait of Cambodia* [Film]. PhotoSynthesis Productions.

Limaye, S. (2019). Unheard voices: Schooling experiences of parents having children with disabilities in India. In M. J. Schuelka, C. J. Johnstone, G.

Thomas, & A. J. Artiles (Eds.), *The SAGE Handbook of diversity and inclusion in education* (pp. 557–570). Sage.

Lok Sabha Passes Bill Allowing Detention Of School Children In Classes 5, 8. (2018, July 18). *NDTV Education Team.* https://www.ndtv.com/education/rte-amendment-bill-2017-lok-sabha-passes-bill-allowing-detention-of-school-children-1885693

Loomba, A. (2015). *Colonialism/Postcolonialism* (3rd ed.). Routledge.

Loomba, A. (2016). The everyday violence of caste. *College Literature, 43*(1), 220–225.

Lynch, R. A. (2011). Foucault's theory of power. In D. Taylor (Ed.), *Michel Foucault: Key concepts* (pp. 13–26). Acumen.

Macaulay, T. B. (2012, April 3 Wednesday). *Minute on Indian Education.* History of English Studies Page. University of California, Santa Barbara. http://www.english.ucsb.edu/faculty/rraley/research/english/macaulay.html

Maddison, A. (2004). *The World Economy, Volume 2: Historical Statistics*. OECD Development Centre.

Majumdar, M., & Mooij, J. (2012). The marks race: India's dominant education regime and new segmentation. In C. Sleeter, S. B. Upadhyay, A. K. Mishra, & S. Kumar (Eds.), *School education, pluralism and marginality: Comparative perspectives* (pp. 207–246). Orient Blackswan.

Manjrekar, N. (2013). Women school teachers in new times: Some preliminary reflections. *Indian Journal of Gender Studies, 20*(2), 335–356.

Maudslay, L. (2014). Inclusive education in Nepal: Assumptions and reality. *Childhood, 21*(3), 418–424. https://doi.org/10.1177/0907568213514778

McDermott, R., Edgar, B., & Scarloss, B. (2011). Global norming. In A. Artiles, E. Kozleski, & F. Waitoller (Eds.), *Inclusive Education in five continents* (pp. 223–235). Harvard Education Press.

McLaren, P., & Farahmandpur, R. (2001a). The globalization of capitalism and the new imperialism: Notes towards a revolutionary critical pedagogy. *Review of Education, Pedagogy & Cultural Studies, 23*, 271–316.

McLaren, P., & Farahmandpur, R. (2001b). Teaching against globalization and the new imperialism: Toward a revolutionary pedagogy. *Journal of Teacher Education, 52*, 136–150.

M East Ward, one of Mumbai's poorest areas, has reported 80 coronavirus positive cases so far. (2020, April 14). *Economic Times.* https://economictimes.indiatimes.com/news/politics-and-nation/as-mumbai-races-to-control-covid-in-dharavi-other-slums-emerge-as-hotspots/articleshow/75117931.cms?from=mdr

Mehrotra, N. (Ed.). (2020a). *Disability Studies in India.* Springer. https://doi.org/10.1007/978-981-15-2616-9_1

Mehrotra, N. (2020b). Situating disability studies: A prolegomenon. In N. Mehrotra (Ed.), *Disability Studies in India* (pp. 1–22). Springer. https://doi.org/10.1007/978-981-15-2616-9_1

Mehrotra, N., & Vaidya, S. (2013). Exploring constructs of intellectual disability and personhood in Haryana and Delhi. In R. Addlakha (Ed.), *Disability studies in India: Global discourses, local realities* (pp. 145–168). Routledge.

Menashy, F., & Manion, C. (2016). The historical evolution and current challenges of the United Nations and global education policy-making. In K. Mundy, A. Green, B. Lingard, & T. Verger (Eds.), *Global policy and policy-making in education* (pp. 319–334). Wiley Blackwell.

Menon, S. S. (2019). *Stakeholder perspectives on teacher attrition in private early childhood schools in India* (Unpublished dissertation). Walden University.

Mercer, J. R. (1973). *Labeling the mentally retarded: Clinical and social system perspectives on the mentally retarded.* University of California Press.

Miles, M. (1981). *A short circuit of mental handicap in India with some sparks of development* [Monograph]. Mission Hospital.

Miles, M. (1985). *Where there is no rehab plan: A critique of WHO's scheme for community based rehabilitation* [Monograph]. Hesperian Foundation.

Miles, S., Lene, D., & Merumeru, L. (2014). Making sense of inclusive education in the Pacific region: Networking as a way forward. *Childhood, 21*(3), 339–353. https://doi.org/10.1177/0907568214524458

Miles, S., & Singal, N. (2010). The Education for All and Inclusive Education debate: Conflict, contradiction or opportunity? *International Journal of Inclusive Education, 14*(1), 1–15.

Mili. (2018). Pedagogical reform in Indian school education: Examining the child-centred approach. *Journal of Philosophy of Education, 52*(3), 533–547.

Ministry of Education. (2018). *Framework for implementation of Samagra Shiksha: An integrated scheme for school education.* Department of School Education and Literacy. http://samagra.mhrd.gov.in/framework.html

Ministry of Human Resource Development. (2019). *Draft National Education Policy 2019.* Government of India.

Ministry of Law and Justice. (2009). *The Right of Children to Free and Compulsory Education Act*. Legislative Department. https://www.mhrd.gov.in/sites/upload_files/mhrd/files/upload_document/rte.pdf

Ministry of Law and Justice. (2016). *The Rights of Persons with Disabilities Act*. Legislative Department. http://legislative.gov.in/actsofparliamentfromtheyear/rights-persons-disabilities-act-2016

Misra, A. (2000). Special education in India: Current status and future directions. *Division of International Special Education Services Journal, 3*, 6–11.

Mitra, S. (2006). The capability approach and disability. *Journal of Disability Policy Studies, 16*(4), 236–247.

Mohan, P. (2017). The road to English: The slow migration of the EWS child to elite India. In M. J. Tiwary, S. Kumar, & A. K. Mitra (Eds.), *Dynamics of inclusive classrooms: Social diversity, inequality and school education in India* (pp. 203–215). Orient BlackSwan.

Mooney, J. (2007). *The short bus: A journey beyond normal.* Henry Holt.

Mosse, D. (2018). Caste and development: Contemporary perspectives on a structure of discrimination and advantage. *World Development, 110*, 422–436. https://doi.org/10.1016/j.worlddev.2018.06.003

Motha, S. (2014). *Race, empire, and English language teaching: Creating responsible and ethical anti-racist practice.* Teachers College Press.

Mukherjee, M., & Kumar, S. (?) Local and global "congeries" shaping educational administration in India. In V. Fernandes & P. W. K. Chan (Eds.), *Asia Pacific education leadership, governance and administration.*

Mukherjee, S. (2016). *The gene: An intimate history.* Scribner.

Mukhopadhyay, R., & Sriprakash, A. (2011). Global frameworks, local contingencies: Policy translations and education development in India. *Compare: A Journal of Comparative and International Education, 41*(3), 311–326.

Mundy, K., & Manion, C. (2015). The education for all initiative. In E. Unterhalter & T. McCowan (Eds.), *Education and international development: Practice, policy and research* (pp. 49–68). Bloomsbury.

Nambissan, G. B. (2012). Private schools for the poor: Business as usual? *Economic & Political Weekly, 47*(41), 51–58.

Naraian, S. (2016). Spatializing student learning to reimagine the "place" of inclusion. *Teachers College Record, 118*(3), 1–46.

Naraian, S. (2019). Diversifying theoretical commitments for a transnational inclusive education: Lessons from India. In N. Singal, S. T. Johansson & P. Lynch (Eds.), *Education and disability in the global South: New perspectives from Africa and Asia* (pp. 107–126). Bloomsbury Academic.

Narayan, J., Thressiakutty, A. T., Haripriya, C., Reddy, K. G., & Sen, N. (2003). *Educating children with learning problems in primary schools: Resource book for teachers.* National Institute for the Mentally Handicapped.

National Council of Educational Research and Training [NCERT]. (1987). *Project Integrated Education for the Disabled: a document.* Author.

National Joint Committee on Learning Disabilities (NJCLD). (2016). *What are LD? National Joint Committee on Learning Disabilities Definition of Learning Disabilities.* NJCLD Online. https://njcld.org/ld-topics/

Nawani, D. (2015). Rethinking assessments in schools. *Economic & Political Weekly, 50*(3), 37–42.

NCERT. (2005). *National Curriculum Framework.* Author.

NCERT. (2012). *What students of Class VIII know and can do: A summary of India's National Achievement Survey, Class VIII, 2012.* Author.

The NEP is the best thing that the Modi government has done since *Swachh Bharat*. (2020, August 23). *The Indian Express*. https://indianexpress.com/article/opinion/columns/education-minister-ramesh-pokhriyal-nep-tavleen-singh-6565927/

Now, a toolkit for dyslexic students in Indian schools. (2012, July 25). *The Times of India*. http://articles.timesofindia.indiatimes.com/2012-07-25/news/32847749_1_tool-kit-dyslexic-english-medium-schools

Omvedt, G. (2003). *Buddhism in India: Challenging Brahmanism and caste*. Sage.

Orellana, M. F., Dorner, L., & Pulido, L. (2003). Accessing the assets: Immigrant youth's work as family translators or "para-phrasers." *Social Problems, 50*, 505–524.

The Pain of Dyslexia, as Told by Bollywood. (2008, June 4). *Washington Post*. http://www.washingtonpost.com/wp-dyn/content/article/2008/06/03/AR2008060303201.html

Pandey, G. (2011, February 15). *An 'English goddess' for India's down-trodden*. BBC News. https://www.bbc.com/news/world-south-asia-12355740

Pandey, J., & Pant, S. K. (2008). The marginalized Dalits: Disadvantages in learning. In K. Thapa & G. M. van der Aalsvoort (Eds.), *Perspectives on learning disabilities in India: Current practices and prospects*. Sage.

Panshikar, A. (2019). Specific learning disability: A hidden disability. In S. Chennat (Ed.), *Disability inclusion and inclusive education* (pp. 175–193). Springer. https://doi.org/10.1007/978-981-15-0524-9_9

Parekh, G. (2017). The tyranny of "ability." *Curriculum Inquiry, 47*(4), 337–343. https://doi.org/10.1080/03626784.2017.1383755

Parker, C. (2010). *Global interactions in the early modern age, 1400–1800*. University Press.

Pawar, U. (2009). *The weave of my life: A Dalit woman's memoir* (M. Pundit, Trans.). Columbia University Press.

Peer, B. (2020, July 31). A friendship, a pandemic, and a death beside the highway. *The New York Times*. https://www.nytimes.com/2020/07/31/opinion/sunday/India-migration-coronavirus.html

Peden, J. (2018). Decolonizing the geography classroom: A call to action for educators to reimagine pedagogy of place. In G. J. Sefa Dei & C. Jaimungal (Eds.), *Indigeneity and decolonial resistance: Alternatives to colonial thinking and practice* (pp. 77–94). Myers Press.

Persha, A. J., & Rao, V. R. P. S. (2003). *Early intervention to intra-uterine growth retardation (IUGR) children at risk for developmental delay*. National Institute for the Mentally Handicapped.

Pritchett, L. (2013). *The rebirth of education: Schooling ain't learning*. Brookings Institution Press.

Pro Bono and Legal Aid Clinic on Education & Legal Services Committee. (2021). *Inclusivity vis-à-vis National Education Policy 2020*. National Law University. https://nludelhi.ac.in/download/publication/Inclusivity%20vis%20a%20vis%20NEP%202020.pdf

Pundir, P. (2020, September 8). *Why does India keep dealing with 'Language Wars'?* https://www.vice.com/en/article/4aymeb/why-does-india-keep-dealing-with-language-wars

Rabukawaqa, E. (2009). Opening address. In P. Puamau & F. Pene (Eds.), *Inclusive Education in the Pacific* (The PRIDE Project; Pacific Education Series No. 6, pp. 1–8). Institute of Education, University of the South Pacific. http://www.usp.ac.fj/fileadmin/files/Institutes/pride/Workplans_and_Reports/Paper_and_publications/IEbook.pdf

Ramaa, S. (2000). Two decades of research on learning disabilities in India. *Dyslexia, 6*, 268–283.

Rao, L. G., Narayan, J., & Mani, M. N. G. (2005). *Status of education of children with disabilities.*

Rao, S., & Kalyanpur, M. (2020). Universal notions of development and disability: Towards whose imagined vision? *Disability and the Global South, 7*(1), 1830–1851.

Rose, R., & Rajanahally, J. (2019). The application of inclusive principles and practice in South India: Successes and challenges. In N. Singal, S. T. Johansson, & P. Lynch (Eds.), *Education and disability in the global South: New perspectives from Africa and Asia* (pp. 165–182). Bloomsbury Academic.

Roy, A. (2019). Foreword. In A. Roy (Ed.), *My seditious heart: Collected nonfiction* (pp. xiii–xxv).

Said, H. (2018). The role of English education in post-colonial Egypt: Criticisms and solutions for the future. In G. J. Dei & C. Jaimungal (Eds.), *Indigeneity and decolonial resistance: Alternatives to colonial thinking and practice* (pp. 125–139). Myers Education Press.

Sainath, P. (2001). *And then there was the market*. (Blog). https://www.india-seminar.com/2001/497/497%20p.%20sainath.htm

Saldaña, J. (2021). *The coding manual for qualitative researchers* (4th ed.). Sage.

Sarangapani, P. M. (2003). Indigenising curriculum: Questions posed by Baiga *vidya*. *Comparative Education, 39*(2), 199–209.

Sarangapani, P. M., & Winch, C. (2010). Tooley, Dixon and Gomathi on private education in Hyderabad: A reply. *Oxford Review of Education, 36*(4), 499–515. https://doi.org/0.1080/03054985.2010.495465

Sarma, P., & Mukhtar, M. (2019, November). *How to Educate One Billion Indians: The Akshar Education Model*. TEDx Talk. https://www.youtube.com/watch?v=IJdAb0mBYWE&feature=share&app=desktop

Sathyamala, C. (2019). Meat-eating in India: Whose food, whose politics, and whose rights? *Policy Futures in Education, 17*(7), 878–891. https://doi.org/10.1177/1478210318780553

Sawhney, S. (2015). Unpacking the nature and practices of inclusive education: The case of two schools in Hyderabad India. *International Journal of Inclusive Education, 19*(9), 887–907. https://doi.org/10.1080/13603116.2015.1015178

Schifrin, N. (2020, May 5). *After 'cutthroat' global competition for PPE, U.S. seeks to make more at home* (audio). PBS Newshour. https://www.pbs.org/newshour/show/after-cutthroat-global-competition-for-ppe-u-s-seeks-to-make-more-at-home

Schwandt, T. A., & Gates, E. F. (2018). Case study methodology. In N. K. Denzin & Y. S. Lincoln (Eds.), *The SAGE handbook of qualitative research* (5th ed., pp. 341–358). Sage.

Scribner, S., & Cole, M. (1973). Cognitive consequences of formal and informal education. *Science, 182*(4112), 553–559.

Sefa Dei, G. J., & Jaimungal, C. (2019). Indigeneity and decolonial resistance: An introduction. In G. J. Dei & C. Jaimungal (Eds.), *Indigeneity and decolonial resistance: Alternatives to colonial thinking and practice* (pp. 1–14). Myers Education Press.

Serpell, R. (2000). Intelligence and culture. In R. J. Sternberg (Ed.), *Handbook of intelligence* (pp. 549–577). Cambridge University Press. https://doi.org/10.1017/CBO9780511807947.026

Shahane, G. (2019, September 14) Why Hindutva supporters love to hate the discredited Aryan Invasion Theory. *Scroll.in.* https://scroll.in/article/937043/why-hindutva-supporters-love-to-hate-the-discredited-aryan-invasion-theory

Shalaby, C. (2017). *Troublemakers: Lessons in freedom from young children in school.* New Press.

Sharan, S. (n.d). *Looking for a shadow teacher? Here's what you should know.* https://buzzingbubs.com/bangalore/understanding-how-shadow-teacher-works

Sharma, R. (2019, February 3). The class divisions that won't fade. *The New York Times*, p. 6.

Shenoy, S. (2016). Efficacy of a bilingual screening tool in L1 Kannada and L2 English to differentiate between language differences and disorders in English language learners. *Psychological Studies, 61*(3), 126–136.

Shenoy, S. (In press). Language difference v social disorder: learning disability and the dyslexia quandary. In *English Language Education and Education for All: social inequality and schooling challenges in the Indian context.* Routledge.

Shenoy, S., & Wagner, R. K. (2019). Language and literacy practices that influence bilingual and bi-literate acquisition in L1 Kannada and L2 English in

Bangalore, India. In R. M. Joshi & C. McBride (Eds.), *Handbook of literacy in Akshara orthography* (pp. 373–388). Springer. https://doi.org/10.1007/978-3-030-05977-4_19
Simpson, M. K. (2007). From savage to citizen: Education, colonialism and idiocy. *British Journal of Sociology of Education, 28*(5), 561–574.
Singal, N. (2005). Mapping the field of inclusive education: A review of the Indian literature. *International Journal of Inclusive Education, 9*, 331–350.
Singal, N. (2006). Inclusive education in India: International concept, national interpretation. *International Journal of Disability, Development and Education, 53*, 351–369.
Singal, N. (2016). Education of children with disabilities in India and Pakistan: Critical analysis of developments in the last 15 years. *Prospects, 46*, 171–183. https://doi.org/10.1007/s11125-016-9383-4
Singal, N. (2019). Challenges and opportunities in efforts towards inclusive education: Reflections from India. *International Journal of Inclusive Education, 23*(7–8), 827–840. https://doi.org/10.1080/13603116.2019.1624845
Singal, N., & Jeffery, R. (2015). Inclusive education in India: The struggle for quality in consonance with equity. In A. J. Artiles, E. B. Kozleski, & F. R. Waitoller (Eds.), *Inclusive education: Examining equity on five continents* (pp. 161–183). Harvard Education Press.
Singal, N., Johansson, S. T., & Lynch, P. (2019). Education of children with disabilities: Changing landscape of new opportunities and challenges. In N. Singal, S. T. Johansson, & P. Lynch (Eds.), *Education and disability in the global South: New perspectives from Africa and Asia* (pp. 1–18). Bloomsbury Academic.
Singh, U. (2009). *A history of ancient and early medieval India: From the Stone Age to the 12th century*. Pearson.
Sinha, U. (1968). The use of Raven's Progressive Matrices test in India. *Indian Educational Review, 3*(1), 75–88.
Sleeter, C. E. (1986). Learning disabilities: The social construction of a special education category. *Exceptional Children, 53*(1), 46–54.
Small, M. L. (2009). "How many cases do I need?": On science and the logic of case selection in fieldbased research. *Ethnography, 10*, 5–38.
Soldatic, K., Melhoe, L., Kermit, P., & Somers, K. (2018). Challenges in global Indigenous-Disability comparative research, or, why nation-state political histories matter. *Disability and the Global South, 5*(2), 1450–1471.
Srivastava, M., de Boer, A., & Pijl, S. J. (2015). Inclusive education in developing countries: A closer look at its implementation in the last 10 years. *Educational Review, 67*(2), 179–195. https://doi.org/10.1080/00131911.2013.847061
Stalin, K. (Director). (2007). *India Untouched: Stories of a People Apart* [Film]. Dhrishti Films. https://www.cultureunplugged.com/documentary/watch-

online/play/5752/India-Untouched-https://www.documentarystorm.com/india-untouched/

Staples, J. (2020). Decolonising disability studies? Developing South Asia-specific approaches to understanding disability. In N. Mehrotra (Ed.), *Disability Studies in India* (pp. 25–41). Springer. https://doi.org/10.1007/978-981-15-2616-9_1

Steiner-Khamsi, G. (2014). Cross-national policy borrowing: Understanding reception and translation. *Asia Pacific Journal of Education, 34*(2), 153–167. https://doi.org/10.1080/02188791.2013.875649

Stiglitz, J. E. (2002, January 12). 'Globalism's Discontents'. *American prospect*. www.globalpolicy.org/globaliz/econ/2002/0114stiglitz.htm. Accessed on November 24, 2004.

Stiglitz, J. E. (2003). *Globalization and its discontents*. Norton.

Stiglitz, J. E. (2007). *Making globalization work: The next steps to social justice*. W.W. Norton.

Stratton, C. (2016). *Education for empire: American schools, race, and the paths of good citizenship*. University of California Press.

Stromquist, N. P., & Monkman, K. (2015). Defining globalization and assessing its implications for knowledge and education, revisited. In N. P. Stromquist & K. Monkman (Eds.), *Globalization and education: Integration and contestation across cultures* (2nd ed., pp. 1–20). Rowman & Littlefield.

Suri, M., & Gupta, S. (2019, March 11). *The land of a million polling stations: India's general election by the numbers*. https://www.cnn.com/2019/02/16/asia/india-election-numbers-intl/index.html

Terzi, L. (2010). *Justice and equality in education: A capability perspective on disability and special educational needs*. Continuum.

Thapa, K. (2008). Learning disabilities: Issues and concerns. In K. Thapa, G. M. van der Aalsvoort, & J. Pandey (Eds.), *Perspectives on learning disabilities in India: Current practices and prospects* (pp. 23–47). Sage.

Thapan, M. (Ed.). (2015). *Education and society: Themes, perspectives and practices* (pp. 428–448). Oxford University Press.

This shadow bridges the gap between child, teacher. (2014, August 6). *Times of India*. http://timesofindia.indiatimes.com/articleshow/39706873.cms?utm_source=contentofinterest&utm_medium=text&utm_campaign=cppst

Tikly, L. (2001). Globalisation and Education in the Postcolonial World: Towards a conceptual framework. *Comparative Education, 37*(2), 151–171.

Timmons, V., & Alur, M. (2004). Transformational learning: A description of how inclusionary practice was accepted in India. *International Journal of Special Education, 19*(1), 38–48.

Tingley, K. (2020, May 17). An international team quickly screened thousands of drugs in search of Covid-19 treatments—An intriguing new model for collaboration in science. *The New York Times Magazine*, pp. 18–19.

Tobin, J. J., Wu, D. Y. H., & Davidson, D. H. (1989). *Preschool in three cultures: Japan, China and the United States*. Yale.

Tooley, J. (2009). *The beautiful tree: A personal journey into how the world's poorest people are educating themselves*. Penguin.

Tooley, J., & Dixon, P. (2005). *Private education is good for the poor: A study of private schools serving the poor in low-income countries*. Cato Institute.

Tyack, D. (2003). *Seeking common ground: Public schools in a diverse world*. Harvard University Press.

UNESCO. (2015). *Fixing the broken promise of Education for All: Findings from the global initiative on out-of-school children*. UNESCO Institute for Statistics.

UNESCO. (2019). *State of the education report for India 2019: Children with disabilities*. Author.

UN News. (2020, April 20). *Onset of COVID-19 pandemic boosts support for international cooperation*. https://news.un.org/en/story/2020/04/1062122

Unni, J. C. (2012). Specific learning disability and the amended "Persons with Disability Act". *Indian Pediatrics, 49*, 445–447. http://medind.nic.in/ibv/t12/i6/ibvt12i6p445.pdf

United Nations Economic and Social Commission for Asia and the Pacific (UNESCAP). (2008). *Training manual for disability statistics*. Author.

Upadhyay, S. B. (2012). Dalits and the modern education in colonial India. In C. Sleeter, S. B. Upadhyay, A. K. Mishra, & S. Kumar (Eds.), *School education, pluralism and marginality: Comparative perspectives* (pp. 152–180). Orient Blackswan.

Urwick, J., & Elliott, J. (2010). International orthodoxy versus national realities: Inclusive schooling and the education of children with disabilities in Lesotho. *Comparative Education, 46*(2), 137–150. https://doi.org/10.1080/03050061003775421

Vagh, S. B., & Sharma, D. (2018). Localized assessments can enhance our understanding of learning at the bottom of the pyramid: Insights from India. In D. A. Wagner, S. Wolf, & R. F. Boruch (Eds.), *Learning at the bottom of the pyramid: Science, measurement, and policy in low-income countries* (pp. 121–137). International Institute for Educational Planning.

Vaidya, S. (2020). Disorder, disability, difference: (Re)presenting autism in India. In N. Mehrotra (Ed.), *Disability Studies in India* (pp. 43–59). Springer. https://doi.org/10.1007/978-981-15-2616-9_1

Vaidyanatha Ayyar, R. V. (2017). Inclusive elementary education in India: The journey. In M. J. Tiwary, S. Kumar, & A. K. Mitra (Eds.), *Dynamics of inclusive classrooms: Social diversity, inequality and school education in India* (pp. 21–43). Orient BlackSwan.

Varma, P. (2007). We need to go back to the drawing board. In A. Kaushik (Ed.), *Shiksha: The challenge of Indian education* (pp. 41–50). Tehelka.

Velaskar, P. (2015). Educational stratification, dominant ideology, and the reproduction of disadvantage in India. In. M. Thapan (Ed.), *Education and society: Themes, perspectives and practices* (pp. 428–448). Oxford University Press.

Vulli, D. (2014). English and medium of instruction: Dalit discourse in Indian education. *Research Journal of Educational Sciences, 2*(2), 1–6.

Waldrop, A. (2015). The meaning of the old school tie: Private schools, admission procedures, and class segmentation. In M. Thapan (Ed.), *Education and society: Themes, perspectives and practices* (pp. 405–427). Oxford University Press.

Wang, Y., Mu, G. M., & Zhang, L. (2018). Chinese inclusive education teachers' agency within temporal-relational contexts. *Teaching and Teacher Education, 61*, 115–123.

Ware, L. (2010). Disability studies in education. In S. Tozer, B. P. Gallegos, A. Henry, M. B. Greiner, & P. G. Price (Eds.), *Handbook of research in the social foundations of education* (pp. 244–260). Routledge.

Why Does India Keep Dealing with 'Language Wars'? (2020, September 8). *Vice News*. https://www.vice.com/en/article/4aymeb/why-does-india-keep-dealing-with-language-wars

Wikipedia. (2020, June 4). *Eighth Schedule to the Constitution of India*. https://en.wikipedia.org/wiki/Eighth_Schedule_to_the_Constitution_of_India

Will, M. C. (1986). Educating children with learning problems: A shared responsibility. *Exceptional Children, 52*, 411–415.

Wilkerson, I. (2020). *Caste: The origins of our discontents*. Random House.

Williams, S. M., Swain, W. A., & Graham, J. A. (2021). Race, climate, and turnover: An examination of the teacher labor market in rural Georgia. *AERA Open*. https://doi.org/10.1177/2332858421995514

Wolfensberger, W. (1972). *The principle of normalization in human services*. National Institute on Mental Retardation.

World Bank. (2020). *India statistics*. https://tradingeconomics.com/india/rural-population-percent-of-total-population-wb-data.html

World Economic Forum. (2019). *India's mega-rich are on the rise*. https://www.weforum.org/agenda/2019/07/india-mega-rich-are-on-rise-inequality/

World Health Organization/World Bank. (2011). *World Report on Disability 2011*. Author.

Worth, J. (2002). *Call the midwife: A memoir of birth, joy and hard times*. Penguin.

Yin, R. K. (2009). *Case study research: Design and methods* (4th ed.). Sage.

Index

M. Kalyanpur, *Development, Education and Learning Disability in India*, Palgrave Studies in Disability and International Development,
https://doi.org/10.1007/978-3-030-83989-5

The manufacturer's authorised representative in the EU is Springer Nature Customer Service Centre GmbH, Europaplatz 3, 69115 Heidelberg, Germany. If you have any concerns regarding our products, please contact ProductSafety@springernature.com

Printed and bound by CPI Group (UK) Ltd, Croydon, CR0 4YY
22/09/2026
02221066-0004